US Marine Corps

Rifle and Pistol Marksmanship

1935

The Old Corps Method of Precision Shooting

Lancer Militaria

Originally published by the United States Marine Corps, 1936

Cover photos: National Archives

Design by Max Crace
Printed in the United States of America

ISBN 0-935856-07-2

Lancer Militaria
P.O. Box 886
Mt. Ida, Arkansas 71957
USA

INDEX

Part I.—RIFLE MARKSMANSHIP

Part I.—GENERAL

1. The purpose of this manual is twofold: First, to present the subject of marksmanship training in such a manner as will be of the most assistance to the inexperienced officer who, through force of circumstances, will have to supervise and carry out the entire training of his company or detachment; second, and to standardize the system of instruction and training for the Marine Corps.

2. The methods herein described have been given a complete test; in fact they are a development of training methods used at the large ranges of the Corps and are considered as being the best yet devised.

3. The manual includes instruction with the .22-caliber rifle, .30-caliber rifle, the pistol, and the subjects related to the training with and firing of such weapons. Although the rifle is the basic weapon for the Marine Corps, yet the exigencies of the service are such that each man should have some training with the pistol, and as many men with the Browning automatic rifle and Thompson submachine gun as the ammunition allowances permit.

RIFLE

4. The object of rifle training, as well as training with the other weapons, is to teach men the use of the rifle so that they will be able to hit the desired object. This ability once attained has an inestimable value upon the morale of the troops engaged in field service.

5. To obtain the best results, individual rifle training is grouped under the following heads:

(A) PRELIMINARY TRAINING—PRESCRIBED YEARLY

6. This must be taught prior to firing a single shot. It includes the teaching of the mechanical technique essential for good shooting consisting of correct positions, aiming, trigger squeeze, use of sling and rapid fire manipulations of the piece; in other words, correct form; also care and cleaning of the rifle and other features which are aids in the making of a rifleman.

(B) .22 CALIBER PRACTICE—PRESCRIBED YEARLY

7. This is an application and continuation of preliminary training. .22-caliber firing adds interest, shows up certain defects which can only be discovered by actual firing, and thus affords the opportunity to correct them, which might not be possible in firing the limited amount of .30-caliber ammunition allowed for training.

(C) INSTRUCTION PRACTICE—PRESCRIBED YEARLY

8. This is the application and continuation of preliminary training in actual firing of the .30-caliber rifle upon the known distance range. Certain defects appear here which have not been previously noticed, and thus an opportunity is afforded to correct them.

(D) RECORD PRACTICE—PRESCRIBED YEARLY

9. Carrying out the test prescribed by **Basic Field Manual,** volume 3, part 1, chapter 1; or **Small Arms Firing Regulations, United States Navy,** to classify the men and determine the proficiency of the organization.

(E) LONG DISTANCE PRACTICE—FIRED WHEN PRACTICABLE

10. Long distance firing practice (800 to 1,000 yards) is advanced training for men qualifying in the higher grades, who are to enter competitions. It develops skill above and beyond that developed by qualification firing, as certain factors enter here which have no place in short-range firing. It is a necessary foundation stone for those who are to be trained as snipers or telescopic sight riflemen.

(F) SNIPING—GIVEN WHEN PRACTICABLE

11. A specialized type of training for only the most expert of riflemen, to make him the equal of a big-game hunter in stalking his prey and defending himself against enemy snipers.

12. As soon after record target practice as possible, training should be given in those features of marksmanship which welds the individual into a perfect military unit for skill, grouped as follows:

(A) MUSKETRY—GIVEN WHEN PRACTICABLE

13. Training in the collective fire of units and its control. The logical sequel to individual training.

(B) COMBAT FIRING

14. The working out of tactical exercises involving the use of service ammunition.

(C) ANTIAIRCRAFT PRACTICE

15. Training to develop individual skill in firing upon rapidly moving targets, and control of such fire.

FINAL RESULTS EXPECTED

16. An organization commander following the course of individual rifle instruction prescribed in this manual should obtain not less than 80 percent qualification as marksmen, sharpshooters and expert riflemen. This figure, as the minimum percentage of qualification to be expected, is not theoretically fixed as a standard, but is based upon the results obtained with thousands of men **and is the lowest figure indicative of conscientious training.** The final result depends upon the thoroughness and conscientiousness with which the preliminary training is carried out, as this preliminary training is the foundation upon which degree of skill in shooting depends and, if carried out in perfunctory, haphazard manner, the desired result will not be obtained. A minimum of 30 hours is required for best results.

17. The proficiency attained by the firer at the first record practice does not remain constant thereafter. "Once an expert, always an expert" is a fallacy, as there is a steady loss in proficiency each month after record practice until the next practice is held. Unless the firer, even if an expert, is given careful training, in his second year, he will usually fail to make the qualification made in the previous practice. This is due entirely to flinching and loss of proficiency in the marksmanship essentials. To illustrate: Some years ago, a large number of expert riflemen were given the record course, preceded by 1 day's preliminary practice of 5 shots at each range to allow them to get their elevations. This test was given to determine the number of men that could be expected to requalify without practice. The result showed that all lost points in direct proportion to the time since last practice. About 25 percent requalified as experts. The remainder made scores equal to sharpshooter and marksmen, and a few failed to make qualifying scores. It was found that men who had fired 2, 3, or more times previously did not show the same percentage of loss, so that it appears, from this test, that after each year's practice, there is a smaller loss in proficiency, probably due to correct shooting habits having been more firmly established.

18. Preliminary practice is followed by .22-caliber practice. If the preliminary training has been conscientiously and intelligently given it will allow the instructor to devote the greater part of his time to overcoming those errors which only show up in actual firing. Also, due to the limited amount of service ammunition available, the .22 firing affords the extra amount of training necessary to become proficient in kneeling and standing. A score of less than 300 is indicative that the marksman will not qualify in his .30-caliber firing.

19. Instruction practice with service ammunition is the final polishing. If preliminary and .22 caliber practices have been properly conducted, it allows the instructors to devote their entire time to those essentials of actual firing which are necessary for a qualified rifleman; to show up the defects in rifles which are found only through actual firing; the ability to fire all shots in rapid fire string; the ability to apply corrections in windage and elevation; determining of the rifle zero; learning to correct from successive rapid fire groups and to center groups on the bullseye and silhouette.

20. The course of instruction for rifle and pistol as given in part III, Schedules, calls for 3 weeks as a **minimum which is desirable when opportunity affords. This applies to all recruits, unqualified and qualified men below the grade of sharpshooter;** also experts and sharpshooters who do not hold two previous consecutive qualifications as expert or sharpshooter. For the experts and sharpshooters not covered in the above, see part III, Schedules. Preliminary and .22 caliber training may be given at the range where men are to fire. In the case of men who are to be sent to another post for their firing, the preliminary and .22 caliber training may be given at the home post just prior to going to the range, thus losing the services of men for a shorter period of time. Commanding officers of permanent ranges, called upon to give the final training to men who received their preliminary training at their home stations, will cause all such men to be examined previous to the beginning of instruction practice to determine whether they are qualified to take up instruction practice, and, if found not so qualified, will hold them 3 weeks, repeating the preliminary training and making report of such cases to the major general commandant.

21. No variations shall be made in the course except as covered in paragraph 20. As previously stated, skill is not maintained without practice and it is true that 90 percent of the men with three average record scores below 301 can be expected to better their scores by taking the full course.

PRELIMINARY TRAINING

22. This training is divided into lessons embracing the subjects to be covered, and given in the order in which they are to be taken up.

23. The first step is the selection and training of assistants (instructors). Select at least 1 instructor for each 7 men of the group who are to fire. This permits of practically individual instruction for each man. An instructor should always have the same men; in this way he soon learns his men and will get better results than if he had a different class each day.

The selection of instructors should be from among the best shots as shown by the service-record books, and preferably from the noncommissioned officers and those who have had previous coaching experience. In making selections bear in mind that these men are to be instructors. Having picked instructors, the next step is to put them through the preliminary training; train them to be letter perfect; these men will be assistants in the training of the company and the final results obtained will depend on the thoroughness of their instruction.

NECESSARY EQUIPMENT

24. In preliminary training the following equipment is necessary:

(*a*) Rifle and sling for each man.

(*b*) Set of shooting pads for each man (issued by Q. M. D. See illustrations 2, 3 and 4. Can improvise, see par. 27).

(*c*) One Marine Corps scorebook for each man (issued by Q. M. D.).

(*d*) Targets A, B, and D (issued by Q. M. D.).

(*e*) Miniature aiming targets (issued by Q. M. D. Can improvise).

(*f*) Shaker aiming device (improvise. See illustrations 5 and 6).

(*g*) Khaki blouse, each man.

(*h*) Pencil, each man.

(*i*) Belgian aiming device. (Good but not absolutely necessary. Issued by Q. M. D. See illustration 7.)

(*j*) Spotters (issued by Q. M. D. Can improvise).

(*k*) Dummy cartridges.

25. Some lessons are preceded by a lecture, the purpose of which is to give the men an insight into the "whys" and "wherefores" of the various steps of training, and thereby obtain better cooperation. This lecture should be given by the senior instructor (company commander), after which each assistant instructor should carry out the remainder of the lesson with his seven men; the senior instructor supervising the work of the smaller groups.

26. The lecture given should embody the salient features to be covered. Those given in this pamphlet are intended as guides and should be enlarged upon and given in the instructor's own words. Each lesson includes a quiz (question and answer). These are given as a guide and should be enlarged upon.

LESSON 1

PADDING BLOUSES; BLACKENING SIGHTS; RIFLE NOMENCLATURE—MATERIAL NEEDED

27. (*a*) Any one of the following articles, the smoke of which is good for blackening sights: Candles, wax tapers, camphor, oily rags, matches, acetylene lamp.

(*b*) Blouse for each man (one size oversize, if possible).

(*c*) Set of pads for each man.

(*d*) Pencil and Marine Corps scorebook, each man.

NOTE.—The Quartermaster's Department issues a standard shoulder and elbow pad. If none are available, pads may be improvised, using sheepskin, felt, turkish towels, etc. Improvised pads should give protection equal to three thicknesses of turkish towel and be of the following dimensions: Elbow 8 by 13 inches; shoulder 7 by 8 inches, with tongue to extend about 3 inches down right arm, protecting muscle.

28. The object of this lesson is to show you how to pad a shooting blouse, to blacken your sights, and to refresh your knowledge of simple rifle nomenclature. When firing on the range, a blouse padded on both the elbows and the shoulder is essential; also wear a woolen undershirt if you have one. This is to protect against sore elbows and shoulders, either of which will seriously affect your shooting. It is possible some of you may have fired in the past without pads, and with no ill results. If so, you are just lucky. Without padding, soreness may develop at any time; perhaps not until just previous to record firing. We will take no chances; **all blouses must be padded.**

29. Sometimes padding is used on a flannel shirt instead of a blouse. This is bad practice. The flannel stretches, allowing the pad to slip from position; and, further, the shirt does not give the proper support to the shooter.

30. Here is a properly padded blouse (**show**). You will notice the top of the shoulder pad is even with the shoulder seam, and the finger of padding extends down the arm to cover the muscle. This covers the spot where the rifle rests against the shoulder. The elbow pads are down nearly to the sleeve cuff and a little to the side of the sleeve. When in the prone position, the sleeve slides up the arm slightly and tends to roll outward, which throws the pads into such a position that the elbows will be in the center of them.

31. A blouse with pads improperly adjusted might just as well not be used; in fact they will do more harm than good. Care must be taken in sewing on the pads that the cloth of the blouse is not bunched or creased under them as this will very likely create blisters.

32. The blouse should be 1 or 2 sizes oversize to allow for the unusual positions used in firing. If an oversize blouse is not available, it may be necessary to rip the back seam; but, if this is done, a gusset must be sewn in. There should be a slight tension across the shoulders when in the prone position.

NOTE.—If improvised, or sheepskin pads of previously given dimensions are used, the following is applicable. The top of the pads should be about 3 inches from the armpit seam. These are so placed to prevent the rifle sling from slipping, and to pad the artery in the arm to keep the pulsation from being communicated to the rifle.

33. In all firing the front and rear sights must be blackened, otherwise they will reflect light and interfere in aiming. When properly blackened, the sights will stand out clearly and well defined against the target. To blacken sights, first clean out the peephole with a toothpick or sharpened match, then remove oil from sights with gasoline. Blacken front and rear of peep sight and the entire front sight, taking care to get a dull coat of black, not mixed with gray streaks.

34. Some knowledge of simple rifle nomenclature is necessary in order that coaches and yourself will speak the same language during rifle training.

35. For the rest of the lesson, assistant instructors take charge of their classes and carry out the following:

(1) Instructor demonstrates prone position as illustrated.

(2) Shows how to mark blouse for padding (illustration no. 1) as follows:

Illustration No. 1

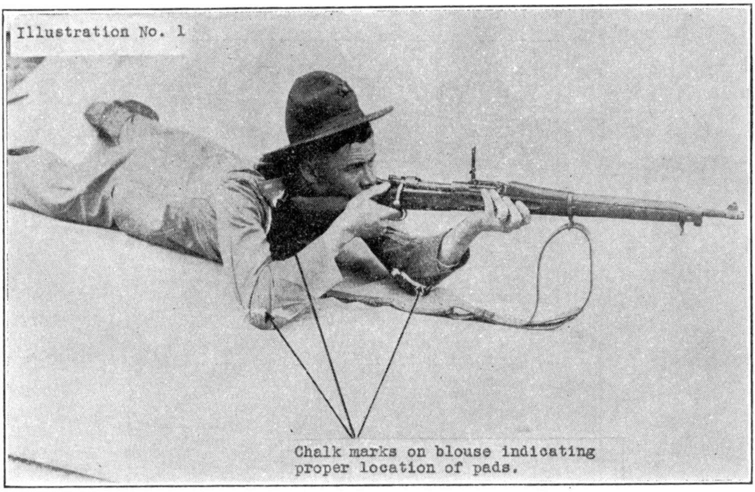

Chalk marks on blouse indicating proper location of pads.

Requires squad to assume the prone position and marks on each man's blouse point where elbows touch ground, and draws line around butt of rifle as it rests against the shoulder. This will indicate spots to be covered by pads.

Has pads temporarily attached to blouses covering the marked places, using safety pins.

Pads having been temporarily attached to blouses, places squad in prone position with rifles and examines position of pads, shifting pads if necessary.

Requires all men to sew the pads on blouses before the next lesson.

The following illustrations show pads properly located 2, 3, 4.

Note.—During the instruction in padding of blouses, the question of a man firing left-handed will frequently arise. Unless he has a previous record of sharpshooter or better, require him to fire right-handed. The rifle was made to fire right-handed; it is awkward to manipulate the bolt left-handed, in fact it is easier to learn to fire right-handed than to learn to overcome the awkward bolt manipulation even though left-handed.

POINTS TO OBSERVE

(*a*) That center of shoulder pads cover spot where rifle rests against shoulder.

Illustration No. 2
Type of Pads issued by Q.M. Dept.

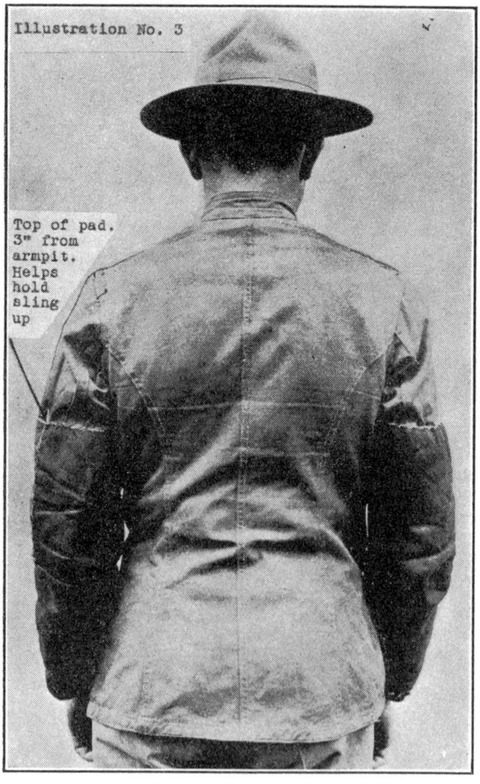
Illustration No. 3
Top of pad.
3" from
armpit.
Helps
hold
sling
up

(*b*) That the shoulder pad extends over and down the right arm, protecting the muscle.

(*c*) That center of elbow pads are over the elbows when in prone position.

(*d*) That the rectangular elbow pads are attached to sleeve 3 inches from armpit seam. This style will help to hold the rifle sling in the proper place on the left arm.

(*e*) That the blouse is not wrinkled under the pads.

3. Demonstrates blackening sights, observing the following points:

(*a*) Clean peephole with match.

(*b*) Remove all grease from the front sight and both sides of the rear sight with gasoline, if available.

(*c*) With smoke, blacken entire front sight and rifle barrel within 3 inches of the sight; blacken the front and rear of the peep sight; with battle sight, blacken the entire top side.

(*d*) Get a dull coat of black without having streaks in it.

(*e*) Wipe black off elevation scale; each side of sight leaf and about one-eighth inch both ends index line on drift slide.

4. Requires men to blacken sights.

5. Issues to each man, pencil and copy of Marine Corps score-book. Has pencils tied to score-books, and cautions men to keep the book for use on the range, and for a reference book thereafter.

6. Holds school on rifle nomenclature. Using a rifle, points out the parts given in the Marine Corps score-book and calls for their names.

7. Quizzes the class:

Q. Why is padding used in shooting?

A. To prevent sore elbows or shoulders while firing.

Q. How do sore elbows or shoulders effect a man's shooting?

A. The soreness makes him gun-shy, which interferes with good shooting.

Q. Why should a blouse be worn instead of a shirt?

A. Because the blouse when buttoned up braces the body, as it does not stretch; thus making it easier to hold steady.

Q. Does a real expert shot ever shoot without pads?

A. Never.

Q. Is a man allowed to shoot left-handed?

A. Only if he has a previous qualification of sharpshooter or expert.

Q. Why is padding required to extend over the upper right arm muscle?

A. In some shooting positions, men are apt to fire the rifle with the butt partially resting against the upper arm muscle.

Q. Why are sights blackened?

A. To help in aiming.

Q. How does a bright sight affect shooting?

A. Reflects the light and makes it hard to see distinctly, thereby causing inaccurate aim.

Q. Why must sights be free of grease or oil before they are blackened?

A. The soot of the smoke will be greasy and will be a shiny black.

Q. Why blacken the back of the sight.

A. If shiny, it will reflect light as well as the front of the sight and hinder in obtaining the best aim.

Q. Should the top of the barrel near the sight be blackened?

A. Yes; this is good practice.

Q. What is the most satisfactory blackening material?

A. Some experts prefer one thing, others another. Matches probably give a thinner more even coat of black, but it is harder to apply.

Q. If a blouse is small, and is ripped, why sew in a gusset?

A. In some firing positions, a certain amount of tautness is desirable across the back to help brace the body muscles.

Q. Could a dungaree coat be used for a shooting blouse?

A. Yes.

Q. Which type of pad do you think would be best, a long one extending well up the arm or the small Q. M. pad?

A. The long one.

Q. Why?

A. It would tend to keep heart pulsation from being communicated to the rifle and help hold up the sling.

Q. On a sandy firing point, would elbow pads be of any help other than to protect the elbows?

A. Yes. To some extent they would act as snowshoes, tending to prevent the elbows' slipping and digging into the sand.

Q. If a pad is sewed on over a creased blouse, what effect will it have?

A. The blouse will form a bunch where creased, and very likely cause soreness.

8. Questions by class.

LESSON 2

SIGHTING AND AIMING—EQUIPMENT USED

36.

(*a*) Shaker Aiming Device. (See illustrations 5 and 6.)

(*b*) Miniature Aiming Targets.

(*c*) Belgian Aiming Device. (See illustration 7.) Not a necessity, but a convenience.

THE SHAKER AIMING SERVICE

Description.—The Shaker device is one whereby two rifles, solidly emplaced, are alined so that the line of aim of both rifles will converge on the same point at a distance of about 50 feet. It is based on the principle of the check telescope on naval guns. It can be constructed from materials at hand and with tools readily obtainable. For details of construction see photographs nos. 5 an 6.

Method of setting up.—Place the box in a convenient location for the aiming exercises, weighting the box down with rocks or sandbags if outdoors, or nailing it to the floor if indoors. Place the two rifles, without slings, in the notches provided, barrels up, muzzles to the front, center of receiver over the rear notch—the line of sight will then converge at a point approximately 50 feet from the box. At this point set up a sheet of paper on which is a black bullseye cut from 50 foot small bore target. Final correction and adjustment of the line of sight of both rifles on the bullseye should be made by means of windgage and rear sight on both rifles. The rifles should be held rigidly in the notches and the box should be held securely or weighted.

First aiming exercise.—The coach having alined the sights of both rifles correctly on the small stationary bullseye at a distance of about

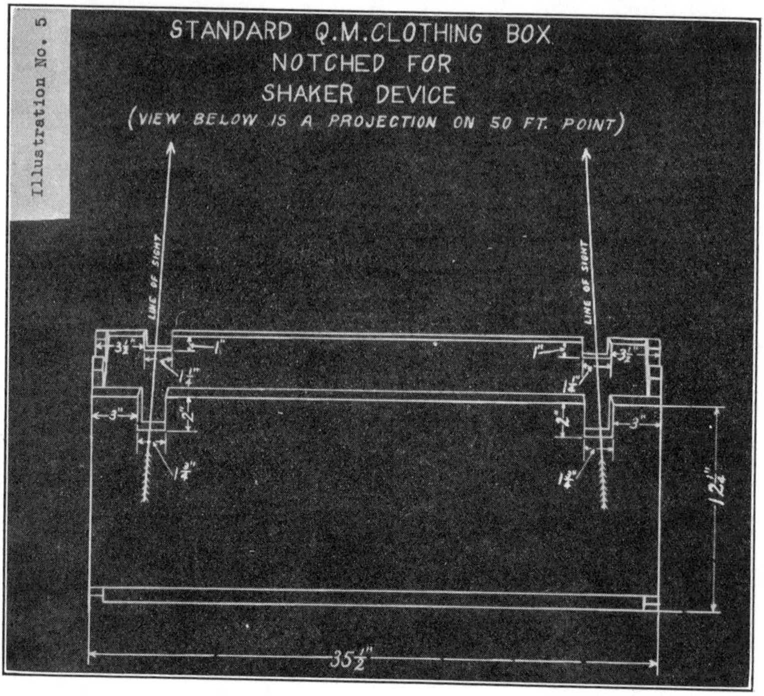

Illustration No. 5

Illustration No. 6

50 feet, takes the prone position at one of the rifles and directs the pupil to take a similar position at the other rifle. The pupil is then directed to look through the sights of his rifle to see the correct alinement of the sights on the bullseye. While doing this the coach explains the correct line of sight.

Second aiming exercise.—Having assured himself that the pupil has seen and understands the correct method of aiming, the coach directs the pupil to look through the sights of the rifle and observe the method by which the coach alines his sights on a movable bulls-eye. An assistant, with a stationary disk similar to that shown in the photograph, is stationed at the target prepared to move the sighting disk as directed. The coach then directs the assistant to place the sighting disk over the stationary bullseye somewhere out of the line of sight, and directs the assistant to move the disk up or down, or to the right or left, until it is properly alined, when he calls, "mark", at which command the assistant holds the disk in that position. The pupil is then asked to explain how the alinement appears to him. If it appears to be correct to the pupil he is directed to repeat the

Bulls-eye.

Target D.

operation just performed by the coach. This exercise is repeated until the pupil is able to aline the sights properly and promptly. If not correct the instructor should check up the alinement of the pupil's rifle and make sure that it has not been moved.

LECTURE BY SENIOR INSTRUCTOR

37. The object of this lesson is to teach the correct manner of aiming a rifle. There is only one way that will give uniform results under all conditions of light. If any other method is used, best results will not be obtained. To shoot accurately, you must know how to aim correctly; good scores will not result from poor aiming. If you do not learn how to aim correctly your shooting will be erratic.

To aim with the peep sight **look through this peep hole** (touch). **See the tip of the front sight** (touch) **in the center of the peep hole; the tip of the front sight should then just touch the bottom of the bullseye; the tip of the front sight always remaining centered in the peep hole.**

1538—36——2

Aiming is simple if you do it in these steps: Look through the peep, center the tip of the front sight in the peep, then direct this line upon the aiming point, focusing the eyes at this step upon the aiming point. After you think you have the correct aim, check on the front sight being centered in the peep. After a little practice, your eyes will automatically center the front sight in the peep. Always aim the same for each shot; do not neglect any step or become careless. Sometimes your vision may become blurred while aiming, if so, look off in the distance for a moment and then try again. Once you have the aim, squeeze the trigger. Nothing is gained by aiming too long; whereas in so doing, your vision may become blurred.

The peep sight should always be used when possible, as it is the most accurate sight.

In range practice, the only sight used is the peep, and in the field it should be used when possible. The rifle is equipped with this battle sight (touch) to be used when there is not time to set the peep sight or darkness prevents use of the peep. In aiming with this sight **look through this notch** (touch). **See the tip of the front sight flush with the top and in the middle of the notch; the tip of the front sight should then touch the aiming point.** (See A.)

In using the open sight (battle sight) at 200 and 300 yards, the point of aim is the center of the object to be hit (waist line of a man). At 500 yards, the point of aim is the top of his hat. These are thumb rules, but will result in your obtaining hits.

The balance of the lesson to be conducted by the assistant instructors.

(1) Explain that in all aiming the cheek should rest against the stock with the eye as close to the cocking piece as the position and build of the man will permit. Resting the cheek against the stock tends to reduce muscle fatigue of the neck and automatically places the eye in correct place for aiming. The nearer the eye is to the rear sight, the easier it is to aim, and, in rapid fire, to recover the aim.

(2) With the sights of the two rifles in the Shaker device both adjusted to give correct aim at a common target (see illustration 6); require each man in turn to assume the proper aiming position and to look through the sights of one rifle, with the instructor at the other rifle, and describe what he sees. Instructors caution man, in conjunction with aiming, to take a half breath before commencing to line sights up and hold it until the trigger is released. If, while aiming, breathing is continued, the body will move up and down as you breathe and it will be impossible to hold the aim steadily on the bullseye. If aim cannot be perfected in one breath, breathe a few times and try again. Do not put lungs to a strain by taking a big breath and holding it; this causes the heart pulsations to be harmful. Hold a small portion of the breath in an easy natural manner.

(3) Repeat paragraph 2 showing battle sight aiming for 300 yards.

(4) Instructor checks each man's knowledge of aim in the following manner:

(*a*) By use of Belgian aiming device if available. See illustration 7.

(*b*) With rifles in the Shaker device adjusted to the same point of aim, station man at the target with a movable hand disk target (see illustration 6), and require each man to aline the target properly.

(5) Repeat the subject with those who do not pass check.

(*a*) When trouble is experienced, sometimes aiming at the instructor's eye (illustration 8) discloses the error being made.

COMMON ERRORS

38. The common errors are:

(*a*) Using the peep, but not the front sight, or vice versa.

(*b*) Careless centering of the front sight in the peep; same is true of the open sight.

(*c*) Not taking half breath, and holding breath.

(*d*) Cheek not against stock.

(*e*) After centering front sight, not focusing eye on target.

(7) Quizzes class:

Q. How many correct ways are there of aiming?

A. One.

Q. When the aim is taken correctly, how should the sights look relative to the target?

A. The tip of the front sight exactly centered in the peep, and just touching the bottom of the bullseye.

Q. Should the tip of the front sight extend into the bullseye?

A. No; it should just touch it.

Q. When possible, what sight should be used?

A. The peep sight.

Q. When should the open sight be used?

A. Only when there is not time to set the peep sight, or light conditions are such that it cannot be used.

Q. After a man has learned to aim, where does he focus his eyes while aiming?

A. He looks at the target through the rear sight and, automatically, his eyes center the tip of the front sight in the rear sight.

Q. When aiming, why should the breath be held?

A. In breathing the body moves slightly, and this disturbs the aim.

Q. In aiming, where should the cheek be?

A. Resting against the stock.

Q. Why against the stock?

A. Because it places the eye in the correct position to aim.

Q. What is the position of the eye relative to the cocking piece?

A. As close to the cocking piece as it can be placed.

Illustration No. 7

Belgian Device in use

Illustration No. 8

Q. What is the advantage of having the eye near the cocking piece?
A. It makes aiming easier and quicker.

NOTE.—Questions involving errors, such as, "What is the effect of using too much front sight?" should be avoided. Only emphasize correct procedure. Men will make mistakes when instructed in the proper methods, and if their minds are further burdened by methods which should not be followed, some will confuse correct and incorrect procedure.

(8) Questions by class.

LESSON 3

SLING ADJUSTMENT AND USE OF SAME

LECTURE BY SENIOR INSTRUCTOR

39. The purpose of the lesson is to teach the adjustment of the sling and the proper method of putting it on. The sling should be

used in all positions; the standing, sitting, kneeling, prone or squatting and when firing with or without a rest. The sling serves two purposes, and both are very important. First: When properly adjusted, it permits part of the recoil of the piece to be taken up by the left hand and arm, thus removing part of the recoil from the shoulder and rendering it less liable to become bruised. It distributes the recoil instead of leaving it concentrated on the one shoulder. (Take prone position and demonstrate as shown in illustration 9.) Secondly, it greatly helps to hold the piece steady, without muscular effort, while aiming. (Demonstrate aiming, prone with and without sling, and have a man place weight on the rifle in each case to show steadiness of hold.) The more muscular effort required, the harder it will be to hold the rifle steady, and the more body tremors will be communicated to the rifle. The sling, properly adjusted, tends to bind those parts of the body used in aiming into a rigid bone brace, requiring less effort than would be necessary if no sling was used. This bracing is practically 100 percent true in the prone position and decreases in the various positions down to the standing, where we find the slightest. Without the sling, accurate shooting is most difficult.

40. Assistant instructors carry out the remainder of the lesson with their classes.

(1) Shows how to shorten and lengthen slings; showing usual adjustment for most men (illustration 10).

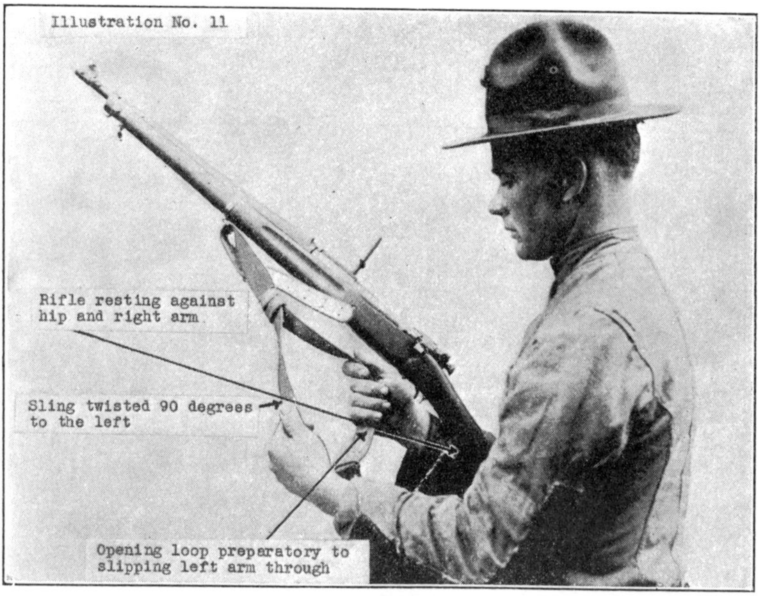

(2) Demonstrates putting on slings (illustrations 11, 12 and 13).

(3) Talk by instructor to his class:

Your sling should be adjusted so that the end of the loop is even with the comb of the stock of your rifle; this is about the average length required. The exact length depends upon the size of the man. With leather slings it varies from the fourth to the seventh hole. Small men require longer slings than large men. When you first begin to use the sling in a shooting position, adjusted as shown, it may seem uncomfortable and too tight, but with a little practice, it will become easy and comfortable. Remember, the sling must be kept tight in order to hold the rifle steady and firm against the shoulder. The sling should be so tight that, when in a shooting position, a strong pressure is exerted on the left hand at the swivel (show) and

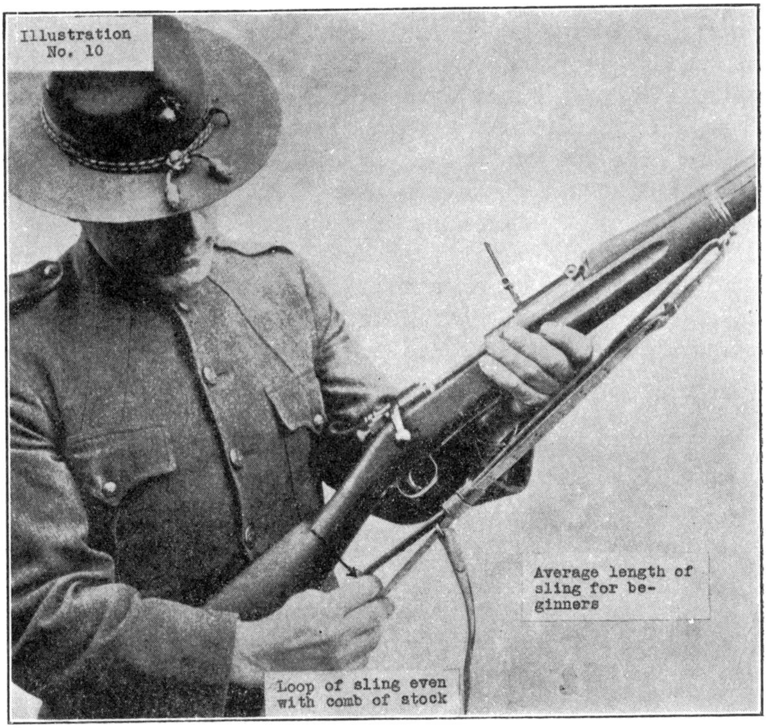
Illustration
No. 10
Average length of
sling for be-
ginners
Loop of sling even
with comb of stock

Illustration No. 12
Push keepers down
tight to hold
sling in place
Sling in
armpit
above
pad

on the left wrist and upper arm (show). It should be of such length that the bolt may be operated without removing the rifle from the shoulder. At first it may seem uncomfortable, but a little practice will make it comfortable, and as it becomes more comfortable you may find that you can shorten it a little. Always keep the sling tight when shooting. Never take in the slack by sliding the left hand back; do it by shortening the sling.

Due to the configuration of the hand, some men get a sore knuckle when their left hand is held against the swivel. This is overcome in some cases by wearing a glove on the left hand. It may be necessary to wrap a handkerchief around the swivel to protect the knuckle. Usually this soreness wears away with the practice.

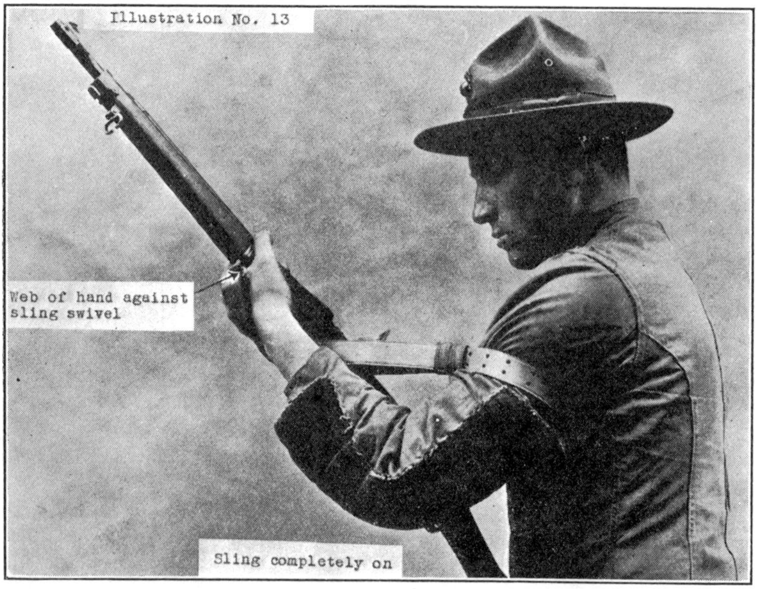

(4) Have the class get into slings. Emphasize the fact that the sling must be kept above the muscle in all shooting and preliminary work. "If it works loose and slips down replace it. Do this for each shot if necessary."

COMMON ERRORS

(*a*) Sling not above arm muscle.

(*b*) Putting left hand through loop from wrong side; resulting in a twisted sling.

(*c*) Metal hook of sling resting against hand or wrist.

(*d*) Rifle resting on fingers of left hand; not on palm.

(*e*) Left hand not carried out against swivel.

(5) Quizzes the class:

Q. What is one advantage of the sling?

A. It takes up part of the recoil.

Q. In what shooting position should it be used?

A. In all positions.

Q. What is the average length and how is it measured?

A. Even with the comb of the rifle stock.

Q. When using the sling, should it be loose or tight?

A. It should be taut.

Q. Where should the sling be on the left arm?

A. Above the arm muscle.

Q. Do expert shots ever shoot without using the sling?

A. Never.

Q. Besides taking up recoil, what does the sling do?

A. Helps to hold the rifle steady.

Q. What does the sling replace?

A. Muscular effort.

Q. In what position does the sling do the greatest good?

A. Standing.

Q. What effect would the metal sling hook have, if it was allowed to rest against the hand?

A. It would bruise the hand in firing.

Q. In the beginning, will a properly adjusted sling seem comfortable?

A. No.

Q. Will it become comfortable with practice?

A. Yes.

Q. If, while in use, the sling seems slack, should it be made taut by drawing the left hand to the rear?

A. No; change the sling length.

(6) Questions by class.

LESSON 4

TRIGGER SQUEEZE

LECTURE BY SENIOR INSTRUCTOR

41. The object of this lesson is to teach the trigger squeeze. In rifle firing the trigger squeeze is very important and demands much practice on your part. A quick pull or yank on the trigger will derange your aim. The trigger must be released without moving the piece, and this can only be done by a slow squeeze on the trigger. Whatever firing position you are in, the second joint of the right forefinger should rest on the trigger, with the right thumb along the stock and the other fingers grasping the small of the stock. (Demonstrate. Illustration 14.)

The grip of the right hand on the small of the stock should be uniform and firm throughout the trigger squeeze but not tense. Take up the slack by one firm pressure of the finger, then, the sights being alined correctly and the hold steady at 6 o'clock on the bullseye, gradually increase the pressure until the rifle fires. The pressure after the slack is taken up must be so steadily applied that the firer will not know when the discharge will take place. Attempting to squeeze the shot off at any instant will result in jerking the trigger and spoiling the aim before the rifle fires. Never increase pressure on the trigger except when alinement of sights and hold are exactly correct. While aiming and executing the trigger squeeze, keep your eyes on the target so that you can tell just where the aim was when

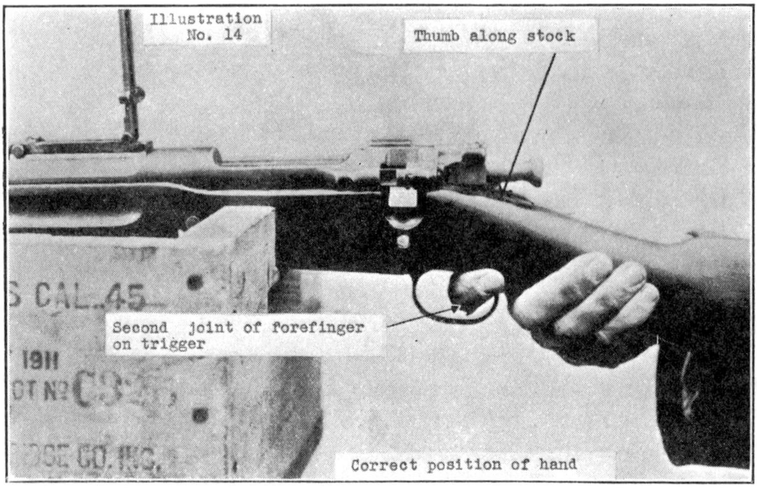

the rifle recoil occurs. This is known in marksmanship as "calling your shot. In firing service ammunition many men, in the beginning instinctively and unconsciously, brace themselves to meet the shock of recoil. This is known as flinching and deranges the aim. By executing a perfect trigger squeeze the bullet has left the rifle before any bracing to meet the recoil can take place, and thus the aim is not deranged. Trying to call the shot is of assistance with the trigger squeeze because the shot cannot be called correctly unless the trigger is squeezed.

No matter how well you aim, you cannot get hits unless you squeeze the trigger. Pulling or yanking the trigger will derange the aim and leads to flinching, for when you yank the trigger you conclude that the rifle will discharge at that moment, and you will brace to meet the recoil, thus spoiling the aim. This applies to rapid fire as well as to slow fire. In rapid fire, you cannot take so much time for the trigger squeeze as in slow fire, but you must squeeze each shot. **The secret of good shooting depends largely upon the trigger squeeze.** One of the principal differences between fair, good, and excellent rifle shots lies in their mastery of the trigger squeeze.

42. Instructors carry out balance of lesson with their classes.

(1) With a cocked rifle demonstrate what is known as **slack.** Explain that the trigger will move a short distance to the rear with just slight pressure applied, then seem to stop, and will move further to the rear if an increased pressure is exerted. The easy motion to the rear is the slack, and with a little practice the slack can be taken up instantly. Once the slack is taken up, the rest of the motion of

Illustration No. 15

releasing the trigger must be accomplished by the trigger squeeze. The reason for taking up this slack instantly is to save the time that would be used in squeezing the slack out of the trigger while aiming. Having taken up the slack, the trigger has only to be moved a very the short distance by squeezing in order to release it.

(2) Using Shaker device, the instructor demonstrates the trigger squeeze. He takes position, aims, takes up slack, and then squeezes the trigger, holding breath and keeping the right eye open.

(3) Placing each man, in turn, at the rifle, the instructor demonstrates the trigger squeeze by placing his finger (illustration 15) over the firer's finger and squeezing the trigger.

(4) Have each man go through the trigger-squeeze exercise until he masters it. Check men's understanding of trigger squeeze by

watching trigger finger and eye, for jerking of finger and blinking of the eye.

COMMON ERRORS

(*a*) Not taking slack out of trigger.
(*b*) Blinking the eye.
(*c*) Jerking or yanking the trigger.
(*d*) Using tip of finger.
(*e*) Not crooking finger around trigger.
(*f*) Using second finger on trigger.
(*g*) Not taking half breath and holding while aiming.
(*h*) Cheek not against stock.
(*i*) Errors in sighting.

(5) Quizzes the class:

Q. What is one difference between an excellent and a poor shot?

A. The excellent shot squeezes the trigger while aiming and only when the hold is steady at 6 o'clock; the poor shot pulls and yanks the trigger, or squeezes the trigger, when the hold is not steady at 6 o'clock.

Illustration No. 16

Q. How does the trigger squeeze help good shooting?

A. The correct trigger squeeze releases the trigger without the shooter's knowledge of when the rifle is going to discharge. When the trigger is pulled or yanked, the shooter knows that the rifle is going off and unconsciously braces himself to meet the shock or recoil, and this "bracing" throws off the aim.

Q. Which is the trigger finger?

A. Right forefinger.

Q. Which joint of the trigger finger is used?

A. The second joint.

Q. Should the right eye be kept open while the trigger is being squeezed?

A. Yes.

Q. What is slack in the trigger?

A. It is that first movement of the trigger accomplished without much finger pressure.

Q. Should the slack be taken out before squeezing the trigger?

A. Yes.

Q. Why?

A. It is a motion which has nothing to do with releasing the trigger, and the time which would be required to take it up by squeezing is saved.

Q. Why is the right eye kept open?

A. To enable the shooter to call the shot.

Q. Of what value is calling the shot?

A. It tends to prevent flinching.

Q. What is flinching?

A. It is tensing of the muscles to meet the shock of recoil before the piece is discharged.

Q. In the trigger squeeze, where is the right thumb?

A. Along the stock.

Q. Where are the other fingers?

A. Grasping the small of the stock.

Q. What is the value of releasing without knowledge of the exact moment it is going to happen?

A. There can be no tensing of muscles, which would derange the aim.

(6) Questions by class.

(7) Examines rifles of class to see that triggers are normal. The trigger should pull not less than 3 pounds trigger weight and once the slack has been taken out the trigger should move smoothly without grating (creep or jump) until it is released. If any rifles do not come up to requirements they should be adjusted by the armorer.

Tighten guard screws and mark position of screw slots by scratch or punch mark on adjacent metal. Explain that loose guard screws will cause shots to be erratic. Check entire front sight for any play and for burrs. Check rear sight. Index line across drift slide should be clear and distinct and exactly horizontal. This line frequently is at an angle and causes many erroneous sight settings. The drift slide must be tight in slide cap in order to keep it in place once it is set. Drift slide binding screw should be checked to see that it holds drift slide assembly in place when it is tightened. The movable base should follow movements of the windage screw and when pressed over by finger pressure should return to original position after pressure is removed. Check bolt. The left locking lug should be free from burrs, as burrs may cause the bolt to stick in rapid fire, cause broken ejectors, and failures to eject empty cartridge. **Do not change bolts unless you have gages to fit the new bolt to the rifle, otherwise you may have a serious accident.** Examine stocks for splits in rear of rear guard screw. (Look on top of stock.) Splits as much as one-fourth inch indicate recoil block is split and guard screws are being driven back into the stock. Most of such rifles will prove to be inaccurate. Split stocks are caused by shooting with loose guard screws.

LESSON 5

PRONE POSITION—EQUIPMENT USED

B TARGETS

NOTE.—In the various positions and rapid fire lessons following regulations targets should be used for aiming. These targets should be the regulation distance away for the kind of fire that is being simulated in conjunction with the position and rapid fire lesson. If the regulation distance cannot be used, reduce the size of the target in proportion to the distance used.

LECTURE BY SENIOR INSTRUCTOR

43. The purpose of this lesson is to teach the prone position. This is the steadiest of all shooting positions. Half of your record shoot-

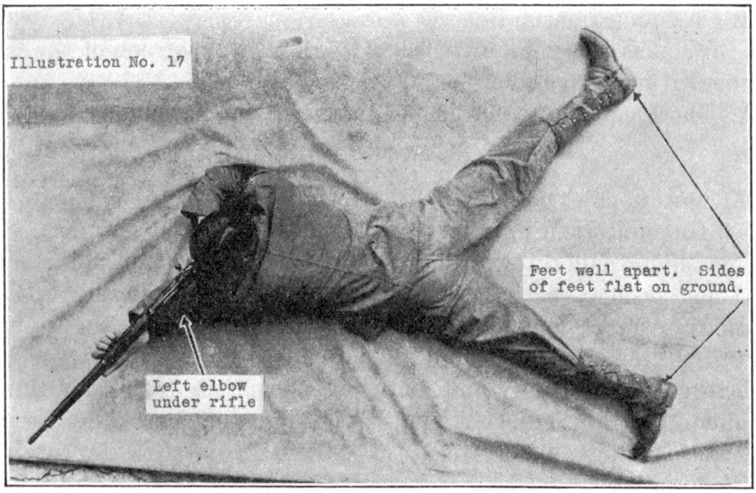

ing will be done from this position, and there is only one correct prone position. At first, the position as taught may seem uncomfortable to you, but with a little practice it will become comfortable and natural. In order to shoot well from this position, it must be comfortable and only practice will make it so. You should not fire a shot until you have become accustomed to this position. (Have an assistant assume the prone position.) Take the prone position by making a half face right and then assume the position. If the rifle, when in position, points to either side of the target, the whole body must be shifted until the rifle naturally points to the target without muscular effort [pointing]. The left hand against the swivel, left elbow under the rifle, right elbow is in such a position that the bolt can be operated, body at a 45° angle to line of fire, feet spread well apart, second joint of forefinger on trigger, thumb along stock, cheek against stock, eye near cocking piece, piece resting in left hand and against shoulder. When this position is properly assumed the correct sling adjustment binds the left upper arm, forearm and right shoulder into a rigid support for the rifle (demonstrate rigidity). In

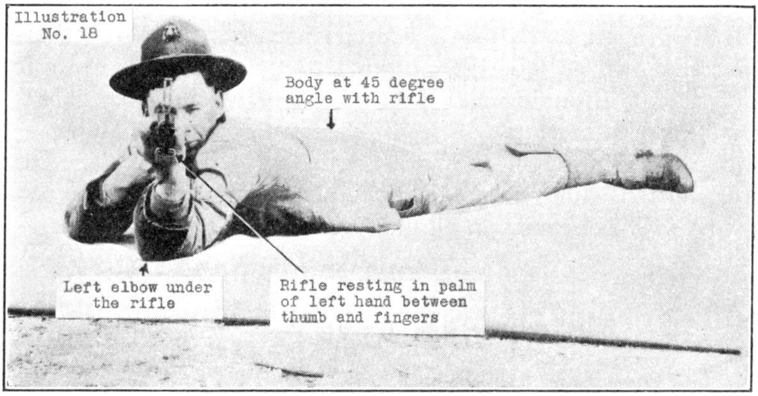

Illustration No. 18

Illustration No. 19

the field, the prone position should always be used when possible. Besides being the steadiest, it presents the least target to the enemy.

44. Instructors carry out balance of lesson with their classes.

(1) Demonstrates prone position as illustrated (illustrations 9, 16, 17, and 18) and explains "In this position, to bring the sights to bear on the target, the muzzle is raised or lowered in either of the two following ways. First, to raise muzzle, with the toes, drag the body

to the rear, keeping the elbows in place. To lower muzzle, reverse this process. Second, to raise the muzzle, move right elbow sideways away from the body. To lower muzzle, reverse this movement. If the sling is too long or slips down the arm, it will cause the left hand to be drawn back in order to raise the muzzle. If the sling is too tight, it will make it difficult to reach the bolt handle."

(2) Requires men to practice prone position, in conjunction with aiming and trigger squeeze, using a B target 500 yards away for an aiming point, dropping rifle from shoulder after each "shot" and going through motions of loading cartridge into chamber. Two 15-minute periods. Check with Belgian device, if available. Lengthens or shortens slings as necessary to permit men to get into position correctly. Checks men on all points while "Snapping in."

COMMON ERRORS

(*a*) Left elbow not under piece.
(*b*) Left hand drawn back under rifle, away from lower swivel.
(*c*) Legs not spread apart.
(*d*) Knees drawn up toward body.
(*e*) Sling too long, or too short.
(*f*) Sling slipped down around elbow.
(*g*) Rifle elbow too far to rear, making it difficult to place finger on trigger and operate bolt.
(*h*) Sights not blackened.
(*i*) Body not at an angle of 45° to the line of fire.
(*j*) Errors of aiming and trigger squeeze.
(*k*) Second joint of trigger finger not on trigger.
(*l*) Tensed or strained muscles causing trembling. Relax.
(*m*) Squeezing trigger when hold is not steady (watch muzzle for movements and unsteadiness).

(3) Quizzes the class:
Q. Where should the left hand be?
A. Against the lower swivel.
Q. Can all men get the left hand out to the swivel?
A. Ninety-eight percent can.
Q. What part of the left hand supports the rifle?
A. The web part between the thumb and fingers and the palm of the hand.
Q. Where should the left elbow be?
A. Directly under the rifle.
Q. What is the position of the legs?
A. Spread well apart.
Q. What should be the angle between the body and firing line?
A. Forty-five degrees.

Q. How is the muzzle raised or lowered to aline the sights on the target?

A. To raise the muzzle: Keeping elbows in position drag body slightly to the rear by pulling on toes or moving right elbow away from body. To depress the muzzle: Reverse this motion.

Q. Where is the right thumb held?

A. Along the stock.

Q. What is the effect of sling properly adjusted and left arm and hand in correct position?

A. They form a rigid triangle for the rifle to rest upon.

Q. What is the steadiest of all shooting positions?

A. The prone.

Q. Where should the head be held?

A. Cheek held firmly against the stock of the rifle, and held at the same place for each shot.

Q. Why the same place each time.

A. If the distance of the right eye from the rear sight varies with different shots, any sighting error present will not be constant, which will tend to scatter the shots. Also, changes in position of hands and body changes the jump of the rifle.

Q. In war time, why should the prone position be used?

A. It presents the least target to the enemy.

Q. If, on assuming the position, the rifle points to one side of the target, should it be slung onto the target by muscular effort?

A. No. The body position must be shifted until the rifle naturally points to the target.

Q. Should the sling hook be allowed to rest against the left forearm?

A. No. It may cause the arm to become sore.

Q. Do you think it possible that the left hand may become sore resting against the lower band?

A. Yes; it may with some shooters.

Q. If this occurs, what can be done; or what should be done to prevent it?

A. Wear a glove on the left hand.

Q. If the sling is too long, what effect does it have?

A. The left hand will be drawn back to support the rifle.

Q. If the sling is too short, what effect does it have?

A. If much too short, the position cannot be assumed. If too short it causes the shooter to assume an incorrect position, varying according to the build of the individual.

Use apropos questions to cover preceding lessons.

(4) Questions by class.

1538—36——3

LESSON 6

SITTING POSITION—EQUIPMENT USED

A TARGETS—D TARGETS

LECTURE BY SENIOR INSTRUCTOR

45. The purpose of this lesson is to teach the sitting position. This position is used in the record course and the greatest steadiness comes only when the various parts of the body have a set place relative to each other, and to the rifle, with correct sling adjustment. The object of practice in the sitting position is to learn to place the parts of the body to get this steadiness. There is more latitude allowed in assuming this position than in the prone position. Three different sitting positions will be shown. Try all three and pick for your use that one which is the steadiest and most comfortable; stick to it, using it whenever you have to shoot in the sitting position. If you continually change, you will never learn any one of them perfectly. Considerable practice is required to become accustomed to this position, and if, when you start actual shooting, you have not become accustomed to it, you will not get the best results (place man in sitting position). The position is assumed by sitting right half-faced to the line of fire. If the rifle does not naturally point at the target shift body position until it does. In whichever sitting position is used, the main essentials are [pointing] left knee under rifle, left hand against swivel, left foreleg in a nearly vertical plane when feet are spread apart, upper arms resting against or on knees, taut sling, cheek against stock, eye near cocking piece, rifle resting in palm of hand. The sitting position ranks next to the prone for steadiness and amount of body exposed to the enemy. This position would be used when the aiming objective cannot be seen from the prone position.

46. Instructors carry out the balance of lesson with their classes.

(1) Demonstrate the three sitting positions as illustrated, cautioning that most men require a sling two holes shorter for the sitting position than used in the prone (illustrations 19, 20, 21, and 22). Explain that the steadiest sitting position for most men is the one with the feet spread well apart. The muzzle of the rifle can be raised or lowered in this position by moving the elbows backward or forward over the knee caps. It can also be done by moving the feet toward or away from the body.

(2) Requires men to practice the three sitting positions, simulating fire at 200 yards using A targets, and to select the steadiest position for their future use.

Caution.—The sitting position is not as steady as the prone and for this reason men are apt to jerk the trigger as the line of aim moves across the bullseye. Although the line of sight may appear to move around the bullseye at 6 o'clock, continue to exert the steady squeeze

Illustration No. 20
Note position of
left elbow

Illustration
No. 21
Same position as
preceding, with
feet together
Left instep against
right heel

on the trigger. If a good aim cannot be taken in one breath, drop the rifle from the shoulder, rest the eyes and try again. Check the aim with Belgian device if available. Give two 15-minute periods.

COMMON ERRORS

(*a*) Point of elbow resting on kneecaps.
(*b*) Left elbow not under rifle.
(*c*) Jerking trigger due to unsteadiness.
(*d*) Body not at about 45° angle to line of fire.
(*e*) Left hand drawn back from swivel.
(*f*) Sling too loose.
(*g*) Sling slipping down arm.
(*h*) Using first joint of trigger finger on trigger.
(*i*) Prolonged aim due to unsteadiness.

(*j*) Errors of aiming.
(*k*) Tensed or strained muscles causing trembling. Relax.

(3) Quizzes the class:

Q. For most men, which is usually the best sitting position?

A. That with the feet well apart.

Q. If the aim cannot be perfected in one breath, what should be done?

A. Rest and try again.

Q. Where should the left elbow be?

A. Directly under the rifle.

Q. Where should the left hand be held?
A. Out against the swivel.
Q. How do you raise or lower the muzzle in this position?
A. By moving the elbows backward or forward over the knees.
Q. Where should the head be held?
A. Cheek against the stock; and against the same part of the stock for each shot.
Q. Do the elbows rest directly on the kneecaps?
A. No.
Q. What really rests against or on the knees?
A. The upper arms.
Q. What is the angle of the body to the line of fire?
A. Approximately 45°, varying slightly with the position used.
Q. Where is the right eye?
A. As close to the cooking piece as possible.

Use apropos questions to cover preceding lessons.

(4) Questions by the class.

LESSON 7

KNEELING POSITION—EQUIPMENT USED

A TARGETS—D TARGETS

LECTURE BY SENIOR INSTRUCTOR

47. The purpose of this lesson is to teach the kneeling position. As in other positions, greatest steadiness comes from a certain placing of the body, with correct sling adjustment, that gives most support to the rifle. This can only be accomplished by considerable practice. To shoot well kneeling, the position must be confortable. Two kneeling positions will be shown. Try each, and select the one which is most comfortable for you. Thereafter, use only that one. The kneeling position ranks next to the sitting position for steadiness. The position is taken by half-facing to the right and dropping to the right knee. (Have man assume correct position.) The essential points to the kneeling position are: Left foreleg nearly vertical, left knee under rifle, right leg about parallel to the firing line, left foot on vertical line midway between right foot and right knee, buttocks resting on inside of right foot or right heel, left elbow extended over left knee, left hand out to swivel, check against stock with right eye near cocking piece, sling taut, body inclined forward. On assuming the position, the rifle should naturally point to the target, if not, shift the position until it does. More of the body is exposed than in the sitting position. It is used when the aiming objective cannot be seen from the prone or sitting positions.

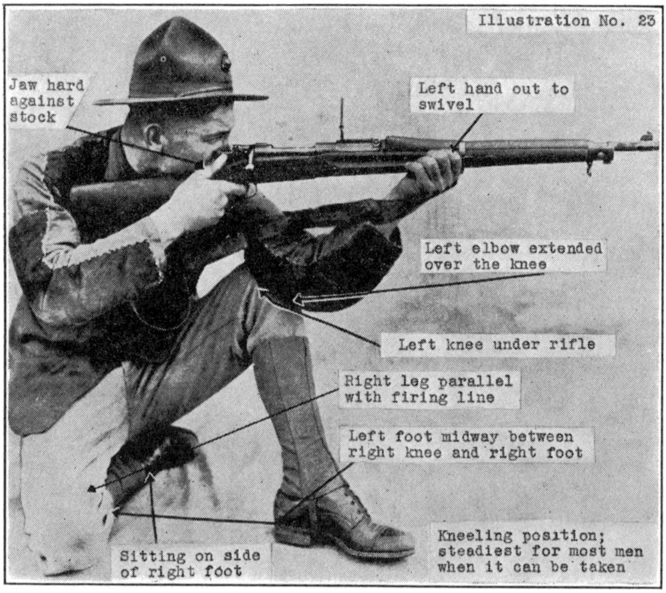

Kneeling position; steadiest for most men when it can be taken

Second best kneeling position

48. Instructors carry out balance of lesson with their classes.

(1) Demonstrate the two kneeling positions as illustrated, cautioning that most men need the same length of sling as in the sitting position (illustrations 23 and 24). The steadiest and most comfortable position for most men is sitting on the heel. The muzzle of the rifle can be elevated or lowered by shifting the weight of the body backward or forward.

(2) Require men to practice the two positions, simulating fire at 200 yards, using an A target, and to select the steadiest position for their future use.

Caution.—The kneeling is not as steady as the prone position, or as the sitting position. Due to this, men are apt to jerk the trigger as the line of aim moves across the bullseye. Although the line of sight may appear to move around the bullseye at 6 o'clock, continue to exert the steady squeeze on the trigger. If good aim cannot be taken on one breath, drop the butt from the shoulder, rest the eyes and try over again. Check aim with Belgian device if available. Give two 15-minute periods.

COMMON ERRORS

(*a*) Body leaning to rear.
(*b*) Left knee not under rifle.
(*c*) Left foot too far forward.
(*d*) Left hand drawn back.
(*e*) Jerking trigger, due to unsteadiness of position.
(*f*) Right leg not parallel with firing line.
(*g*) Point of elbow resting on knee cap.
(*h*) Body not at 45° angle to firing line.
(*i*) Sling slipping down arm.
(*j*) Prolonged aim, due to unsteadiness of position.
(*k*) Sling too loose.
(*l*) Tensed or strained muscles causing trembling. Relax.

(3) Quizzes class:

Q. Which is the best kneeling position for most men?

A. Sitting on heel of foot.

Q. Where should the left elbow be?

A. Directly under the rifle.

Q. What part of the elbow rests on the knee?

A. The elbow joint should be over the kneecap; the flat part of the arm behind the elbow joint should rest on the knee.

Q. On assuming the position, if the rifle does not point at the target what should be done?

A. Shift the body until it does.

Q. How should the muzzle of the rifle be raised or lowered?

A. By shifting the weight of the body forward or backward.

Q. What is the position of the left foreleg?

A. Nearly vertical.

Q. Is this as steady a position as prone or sitting?

A. No.

Q. If your aim moves around or across the bullseye, should you pull the trigger as it crosses the bullseye?

A. No. Never fire except by a squeeze.

Q. What is the position of the right leg?

A. Parallel with the firing line.

Use apropos questions to cover preceding lessons.

(4) Questions by the class.

(5) One 15-minute period simulating fire in sitting position at 300 yards.

LESSON 8

PRONE WITH SANDBAG REST—EQUIPMENT USED

FILLED SANDBAGS—A AND B TARGETS

LECTURE BY SENIOR INSTRUCTOR

49. We now take up the prone position, using a sandbag rest for the rifle. This is a position required in firing for record, and, although

you have had practice in the prone position, resting the hand on a sandbag in this position presents additional difficulties. To meet and overcome these difficulties is the object of practice in using the sandbag position. To most men, the sandbag position is very uncomfortable at first. Good shooting here depends upon being in a comfortable position; and this rests upon having the bag of proper height and sling of correct length for each individual. Shooting from a rest in the prone position is a common battle practice.

50. Instructors carry out balance of lesson with their classes.

(1) Demonstrates position as illustrated, cautioning that the sling should be two holes longer, for most men, used in the prone position (illustration 25), explains that the bag should be built up or flat-

tened down, as necessary, to allow rifle being pointed at target. The best position is with only the back of the hand resting on the bag. In this position the body should be at a 15° to 20° angle to the rifle. The left elbow need not be directly under the rifle. The same pressure should be on the bag for each shot and the same part of the hand should rest against the same part of the bag for each shot.

(2) Requires men to assume position, simulating fire, 600 yards, B target, two 15-minute periods.

COMMON ERRORS

(*a*) Rifle resting on bag.

(*b*) Bag too low or too high.

(*c*) Sling too tight.

(*d*) Errors noted previously: Not holding breath, aiming, jerking trigger, first joint used on trigger, tensed muscles.

(3) Quizzes class:

Q. What part of the hand rests on the bag?

A. Back of the hand.

Q. What length of sling should be used?

A. Each man must learn the length to fit him, usually two holes longer than used in prone.

Q. What are the main essentials to the sandbag position?

A. Correct height of bag and length of sling.

Q. Why should the back of the hand rest on the bag and not the rifle?

A. This is more comfortable and just as steady.

Q. Is the left elbow held directly under the rifle?

A. Not necessarily.

Q. Should the left hand be kept out to the swivel in this position?

A. Yes.

Q. Why?

A. To help take up the recoil and hold the rifle steady.

Q. Is the body at a 45° angle to the rifle?

A. No; only a 15° to 20° angle.

Q. How is the muzzle of the rifle elevated or lowered?

A. By adjusting the sandbag.

Q. Can it be done otherwise?

A. Yes; by hunching the body to the rear or forward.

Q. Between shots, is it permissible to change position on the bag?

A. No. Each shot should be fired with the same part of the hand touching the same part of the bag, in order to have same pressure on bag for each shot.

Q. Why do you think it good practice to have the same pressure on the bag for each shot?

A. In order that the same recoil effect may be had for each shot.

Q. Would the effect of recoil be different if one shot was fired with the rifle resting on the bag, and the next resting the hand on the bag?

A. Yes.

Q. What would be the result?

A. There would be a variation in the bullet strike.

Use appropos questions to cover preceding lessons.

(4) Questions by class.

(5) Simulate firing: 5 minutes, 300 yards sitting, and 10 minutes, 300 yards kneeling.

LESSON 9

STANDING—EQUIPMENT USED

A TARGETS

LECTURE BY SENIOR INSTRUCTOR

51. The purpose of this lesson is to teach the standing position and the sling adjustment used. The standing position is the most unsteady of all firing positions, and naturally so for there is no rest for either arm; the left upper arm must be free from the body [demonstrating]; right arm has no rest, the piece being held in position largely through muscular effort. The only assistance given the body is through proper use of the sling, which tends to hold down the effect of body and muscular vibration in supporting the rifle. The firing regulations prescribe that only the "hasty" sling (illustration 26) can be used in this position. Heretofore you have been using the "loop" sling. In the hasty sling, the lower loop is let out to form a bight, which for most men should be about 1 inch below the butt when the piece is held vertical (demonstrate). To put on the hasty sling, the rifle is held nearly vertical in the right hand, the sling is given a half twist to the left, the left hand and arm are passed between the sling and rifle, the left arm is moved outward bringing the sling high up on the arm, the left hand is passed under the sling, then over the sling, regrasping the rifle. This places the sling along the right-hand side of the hand and wrist. The essential points of the position are [point out]: Body erect, well balanced and facing 45° or more from line of fire, feet spread apart 12 or more inches, left elbow well under the piece, rifle resting on left hand, which is under or slightly forward of the rear sight, piece well up in right shoulder, right elbow approximately at height of shoulder, cheek against the stock, head well forward without strain, sling taut. The same position may be used without a sling, but, except for experienced marksmen, it is advisable to use the sling. The position itself is not so uncomfortable as are some of the others when first starting out; it is, however, much more unsteady. The main thing to learn in this position is to hold the aim just long enough for a quick trigger squeeze. It takes practice. In the field this is the poorest position, and should only be used when no other position is available.

52. Instructors carry out balance of lesson with their classes.

(1) Demonstrates the standing position (illustrations 26, 27, and 28).

(2) Require the class to assume the position and experiment with feet at different distances apart and left hand under and forward of rear sight to determine which seems the most steady for each individual shooter. Check and readjust slings.

Illustration No. 26

STANDING POSITION

Showing hasty sling adjustment, position of rifle in shoulder and cheek on stock.

(3) Impress upon class: Remember—and mark in your score books—your sling length, approximate position of your feet, where you hold your left hand in order that you may always assume this position in the same way. In conjunction with snapping in and firing in this position, you will find it the most unsteady of any you have had. In this position, you may find that by pulling the rifle against the shoulder with a slight pressure of the right hand you will have a steadier hold. Experiment, using slight pressure of both hands. Adopt that method which seems the steadiest to you. Due to the unsteadiness of this position, you may become impatient and

jerk the trigger as the line of aim comes in contact with the bullseye. This will never do. It may happen on some shots, that, while aiming, the rifle will steady down on the bullseye for a short interval. Do not jerk the trigger; try to use a little faster squeeze. Don't aim too long; take a rest and try again. If the rifle never seems to stop long enough on the bullseye to squeeze off the trigger, fire in this fashion: Get as good an aim as possible; if the line of sight moves across the bullseye, even if from 3 ring to 3 ring, and it is the best you can do, carry through a perfect trigger squeeze during this movement. This movement, when combined with a perfect trigger squeeze

will give you an impressive score; you will have 3s, 4s, and 5s, making a good average. If you use anything other than a perfect trigger squeeze, you will have 0s and 2s in your score. Yanking the trigger leads to flinching. Patience and extreme care in squeezing the trigger are highly necessary in the standing position. Never try to hurry in firing in this position.

(4) Give the class two 15-minute periods, simulating fire (snapping in), "A" target at 100 yards.

NOTE.—"A" target at 100 yards is for morale effect, first time using this position. When firing in the standing position encourage men to sit down between shots, using ammunition boxes or camp stools. Continual standing tends to make men nervous.

COMMON ERRORS

(*a*) Left elbow not under piece.
(*b*) Feet too close together.
(*c*) Body faced too much toward target.
(*d*) Body inclined to front or rear; not erect and natural.
(*e*) Right elbow too low.

(*f*) Piece low on shoulder.
(*g*) Snapshooting; jerking trigger.
(*h*) Aiming too long.
(*i*) Head forced to front; straining neck.
(*j*) Sling too loose.
(*k*) Tip of finger on trigger.
(*l*) Left hand held in rear of front sight.
(*m*) Left upper arm resting against body.
(*n*) Tensed or strained muscles causing trembling. **Relax.**

(5) Quizzes the class:

Q. What sling is permitted in this position?

A. Hasty sling.

Q. What is about normal hasty sling length?

A. Bight of sling about 1 inch below butt when rifle is held vertically.

Q. Is it permitted to pass the left arm through the loop of top strap of the sling?

A. No.

Q. Can the left arm be held against the body?

A. No.

Q. May the left hand be held in rear of the rear sight?

A. No.

Q. What angle should the body be in relation to the line of fire?

A. Faced away, at least 45°.

Q. How far apart should the feet be?

A. At least 12 inches.

Q. Should each individual have his feet the same distance apart?

A. No. He must determine what is best for himself.

Q. How should the body be held?

A. Erect and natural.

Q. How is the head held?

A. As far forward as possible without strain.

Q. Where should the left elbow be placed?

A. Under the piece.

Q. Where should the left hand be held?

A. Under and slightly forward of the rear sight.

Q. Does the sling help in this position?

A. Yes.

Q. Which is the least steady position?

A. Standing.

Q. In the field, when should this position be used?

A. Only when no other position can be used.

Q. In this position, the line of sight will move around for most men. Is it proper to pull the trigger when the line of sight touches the bullseye?

A. No.

Q. If the line of sight never steadies down on the bullseye, should the trigger squeeze still be used?

A. Absolutely.

Q. Suppose the trigger squeeze is not used, what will be the result?

A. A much poorer score than if it had been used.

Q. Why?

A. Any other trigger pull leads to flinching.

Q. What is the position of the rifle butt?

A. High on the shoulder.

Q. Why is this done?

A. So that the head may be held in a more nearly natural position than would be possible with the rifle held lower down.

Q. Is pulling the rifle against the shoulder of any help?

A. To some it is.

Use apropos questions to cover preceding lessons.

(6) Questions by class.

(7) Simulate fire: 10 minutes, 600 yards, sandbag rest; 5 minutes, 300 yards, kneeling.

LESSON 10

FIVE HUNDRED YARDS RAPID FIRE—EQUIPMENT USED

DUMMY CARTRIDGES—D TARGETS

LECTURE BY SENIOR INSTRUCTOR

53. The subject of this lesson is rapid fire at 500 yards. Practically half of the record course is rapid fire. Good rapid-fire scores depend upon getting the correct position and aim quickly, fast trigger squeeze for each shot, keeping the butt in the shoulder when operating the bolt, smooth and fast manipulation of the bolt in loading from the magazine, keeping the eye on the target when operating the bolt, and correctly inserting the new clip. This combination of things to do in firing 10 shots in a limited time means considerable practice in simulated fire to acquire the necessary facility and form to get good results. The time allowed for 10 shots is ample for good shooting but it requires speed in loading to give time for squeezing each shot off. Each shot must be squeezed off. In rapid fire the trigger squeeze must be faster than for slow fire; nevertheless it must be a squeeze. In the prone position the bolt must be operated without removing the rifle from the shoulder or raising the right elbow from the ground. This saves time for aiming and trigger squeeze. The new clip must be properly loaded into the magazine, otherwise it may break causing loss of valuable time [demonstrate]. The clip is placed in the magazine so that the thumb rests over the powder space of top cartridge, near the clip, with fingers against the floor plate. The cartridges are shoved into the magazine with a firm, steady pressure. The position is essentially the same as for prone slow fire. The right elbow must be so placed that the right thumb and forefinger can grasp the bolt handle and operate the bolt. The most important shot in a rapid fire string is the first; get it off deliberately with a good trigger squeeze. This tends to steady the firer, and squeezing the first shot will help set up the habit to squeeze each of the others. The firer is apt to be nervous about this time and unless he guards against it will very likely jerk the trigger.

NOTE.—All simulated fire should be carried out just as if it was real firing; using commands, if necessary, to simulate appearance and disappearance of targets. Dummy cartridges should be used in rapid-fire practice. They are useful in learning to load the magazine, provide the same resistance in operating the bolt, and require the same bolt operation as live cartridges. Whenever possible, simulated rapid-fire practice should be carried out with the normal distances with D targets, using target carriers if possible. Time of target exposure should be the same as the prescribed time.

54. Instructors carry out balance of lesson with their classes.

(1) Explains: In prone rapid fire, some men require a sling 1 or 2 holes longer than required in prone slow fire. This is usually due to a slightly different position required. A steady, comfortable slow-fire prone position for some is such that the same position does not permit of ready bolt action and, consequently, has to be varied in rapid

fire. This is accomplished by lengthening the sling to permit the right elbow to be placed farther forward in order to operate the bolt. If the sling is too short in this position, it makes trouble in reaching and operating the bolt; if too long, the firer will draw the left hand back on the rifle to tighten the sling and retain the butt against the shoulder. In rapid fire, trouble is frequently experienced due to too little or too much oil on the bolt. For smooth bolt action, a slight film of oil should always be kept on the working parts. A slight film of cup grease should be used on the cocking cam and on the extracting cam. Grease should be applied to those cams daily during range practice. Too much oil should not be used on the bolt as it may result in splattering oil in the eye.

(2) Assume prone rapid-fire position and demonstrate loading from the magazine (illustrations 29, 30, 31). Explain: Keeping piece against shoulder, right elbow on ground, open bolt, at same time

lowering muzzle to right, draw bolt fully to rear, load new cartridge, at same time bring muzzle back to position, turn bolt handle fully down, eyes on target during entire movement. Demonstrate reloading magazine.

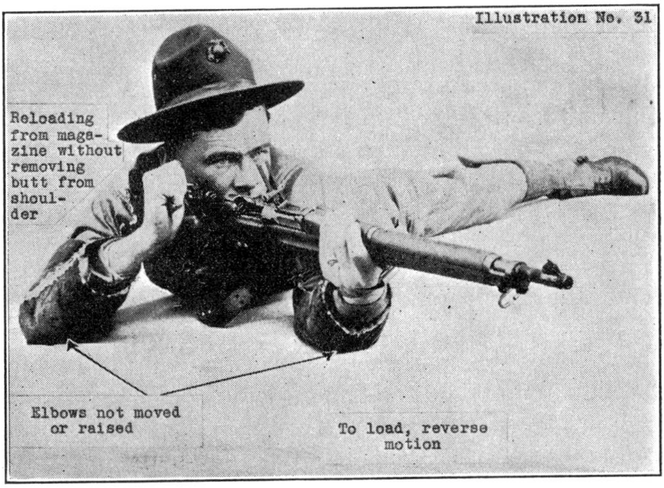

(3) Requires class to simulate four strings, 500 yards rapid fire, using dummy cartridges.

Note for Instructor.—Some men cannot place the right elbow far enough forward to operate the bolt properly when in the prone position. This can be corrected by loosening the sling one hole at a time, keeping the left elbow in place, and shifting the body until it is more nearly in line with the rifle.

1538—36——4

COMMON ERRORS

(*a*) Lifting right elbow off ground to operate bolt.
(*b*) Taking butt from shoulder to operate bolt.
(*c*) Taking eyes off target while working bolt.
(*d*) Not grasping bolt with thumb and forefinger to manipulate it.
(*e*) Right elbow too far to rear, interfering with bolt manipulation.
(*f*) Sling too short, or too long.
(*g*) Muzzle not tilted to right as bolt is opened.
(*h*) Bolt not drawn fully to rear when loading.
(*i*) Errors of position, aiming, etc.

(4) Quizzes the class:

Q. When working the bolt, where should the eyes be looking?

A. At the target.

Q. Should the rifle be removed from the shoulder?

A. No.

Q. What part of the hand should grasp the bolt in working it?

A. The right thumb and forefinger (demonstrate).

Q. What are the essentials for a good rapid fire score?

A. Proper position, fast bolt manipulation, squeezing the trigger, getting the aim quickly, fire all 10 shots, 10 good hits.

Q. If the sling is too tight in rapid fire, what will happen?

A. Usually interferes with the bolt action.

Q. If too loose?

A. Causes firer to draw left hand to rear.

Q. What is the difference between prone slow and rapid fire positions?

A. None, except that the sling may be a little longer for rapid fire.

Q. If the bolt works hard what should be done?

A. Put a little oil on the working parts, grease on cocking and extracting cams.

Q. Is there any difference between the trigger squeeze used in rapid and slow fire?

A. Yes. In rapid fire it should be a faster squeeze, but never a jerk. It is always a squeeze.

Q. Why should the rifle be reloaded quickly?

A. To give more time for aiming and trigger squeeze.

Q. Why should the eyes be kept on the target while reloading?

A. So as not to lose it, and to save time catching aim.

Q. Why should the bolt be drawn fully to the rear?

A. To pick up the next cartridge from the magazine.

Q. Why should the bolt handle be turned fully down upon reloading?

A. If the bolt is not fully closed, the striker will not hit the primer of the cartridge hard enough to fire it.

Q. How should the clip be loaded into the magazine?

A. Place the clip in the clip slot, put the thumb over the rear end of the top cartridge, rest of fingers on magazine floor plate and squeeze the hand together.

Q. Is the piece taken from the shoulder to load it?

A. No.

Q. Is the right elbow removed from the ground in reloading?

A. No.

Q. Due to shortness of time, is it permissible to yank the trigger?

A. No.

Q. Why not?

A. It deranges the aim and leads to flinching.

Q. Which is the short on which to exercise the most care?

A. The first.

Q. Why?

A. The use of care in getting a good trigger squeeze tends to steady the firer and tends to establish the habit of squeezing each shot in the string.

Use apropos questions to cover preceding lessons.

(5) Questions by class.

(6) Give three 10-minute periods simulated fire. One for sitting (slow fire) 300 yards; one for kneeling (slow fire) 300 yards, and one for prone, sandbag (slow fire), 600 yards.

LESSON 11

THREE HUNDRED YARDS RAPID FIRE—EQUIPMENT USED

DUMMY CARTRIDGES—D TARGETS

LECTURE BY SENIOR INSTRUCTOR

55. The subject of this lesson is rapid fire at 300 yards. This not only involves the bolt manipulation, etc., used in 500 yards rapid fire, but involves dropping from the standing position into the prone position, after the targets appear. They are exposed 10 seconds less for this range than for the 500-yard range. This means practice in going from the standing to the prone position and a speeding up of all the other operations, in order to save time for the actual shooting, and to get off 10 well-aimed shots in the minute and 10 seconds allowed. The time limit is ample for well-trained men, but to reach this stage requires considerable preliminary practice. Demonstrate prone position from standing (illustrations 32a 32, 33, 34, 35). The rifleman deliberately assumes the prone position, marking spots where right and left elbows are to rest; he then arises, without moving the toes from the position occupied when lying prone, and awaits the appearance of the targets. When the targets appear, he, holding the rifle in the left hand, without moving the feet, drops upon his knees and simultaneously allows his upper body to fall forward, breaking the fall with the free right hand, and drops the left elbow into the spot previously marked. He then places the butt of the rifle in the shoulder, grasps the stock and places the right elbow in position.

SEQUENCE OF ASSUMING PRONE POSITION RAPIDLY

Ready.—Position of elbows previously located, feet in place for assuming prone position as previously located.

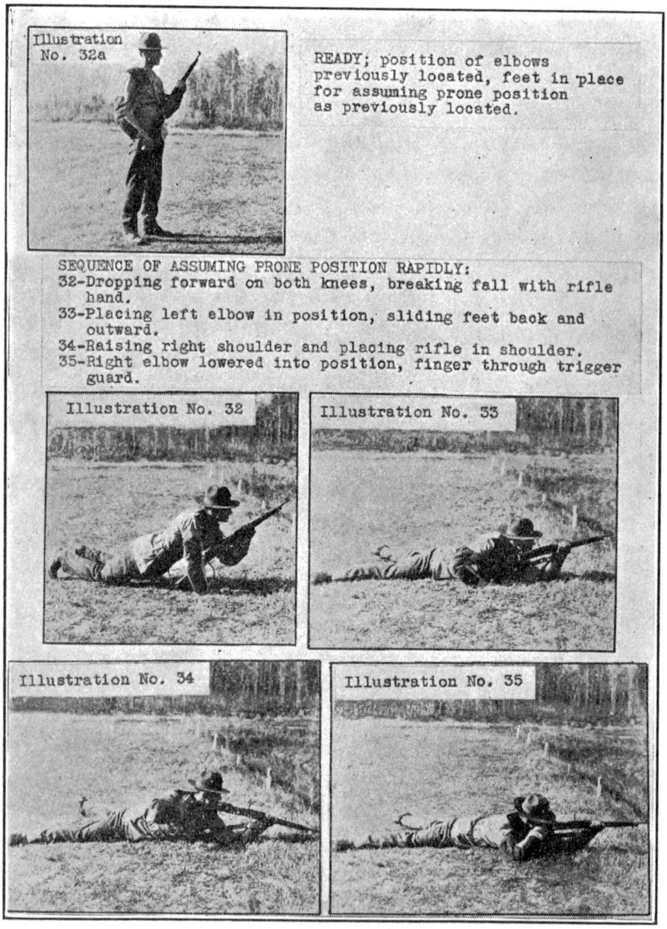

Dropping forward on both knees, breaking fall with right hand.
Place left elbow in position, sliding feet back and outward.
Raising right shoulder and placing rifle in shoulder.
Right elbow lowered into position, finger through trigger guard.

56. Instructors carry out balance of lesson with their classes.

(1) **Cautions the class.**—When a man goes on the line for a string of 300 yards rapid fire, he should assume the prone position and locate the spots where his elbows rest, and then stand in such a position, while awaiting the appearance of the targets, that when he drops into the prone position his elbows will fall into the selected spots without having to hitch his body around (demonstrated).

(2) Give four strings, simulated rapid fire, 300 yards.

COMMON ERRORS

(*a*) Not locating elbows previous to getting into position to shoot.

(*b*) The errors common to 500 yards rapid fire.

(3) Quizzes class:

Q. Why take the prone position previous to a string of 300 rapid fire?

A. In all shooting a comfortable position is of prime importance. By assuming the prone position before the rapid-fire string is run, the shooter is enabled to locate a comfortable place on the firing line and can fix the position of his elbows so that when he drops down he will find himself in a desirable position and will not be worried by having to shoot from an uncomfortable position, or lose valuable time in changing positions.

(4) Questions by class.

(5) One 10-minute period, simulate slow fire, 200 yards standing.

LESSON 12

200 YARDS RAPID-FIRE DIMENSIONS OF TARGETS—EQUIPMENT USED

DUMMY CARTRIDGES—A AND D TARGETS

LECTURE BY SENIOR INSTRUCTOR

57. The subjects to be covered in this lesson are rapid fire at 200 yards and target dimensions. Rapid fire at 200 yards not only involves good rapid-fire form, but necessitates practice in dropping from the standing to the sitting position, after the targets appear, and the firing of 10 shots; all to be done in 1 minute. This can be done, and 10 well-aimed shots fired; but it means practice in quickly assuming a comfortable sitting position, and pratice to attain speed and smoothness in the other essentials of this form of fire, in this position. None of the slow-fire principles should be slighted, but they must be carried out faster in rapid fire. In record rapid fire, nothing but the sitting position will be used at this range.

A knowledge of target dimensions is of importance to a shooter, in order that from a marked hit on a target he can figure exactly how far his shot is from the center of the bullseye. The knowledge of this distance is necessary in calculating the correct amount of change to be made in the sights to place the next shot in the bullseye. This (point out) A target is 4 by 6 feet with a 10-inch bullseye, and, approximately, 9-inch rings. This (point out) B target is 6 by 6 feet with a 20-inch bullseye and 9-inch rings. Remember three things, 10-inch bullseye for the A target; 20-inch bullseye for the B target, and approximately 9-inch rings. With this knowledge, you can locate your hits on either of these targets. This (point out) is the D or rapid-fire target. As you can see from looking at this target that it is not of

Illustration No. 36

Showing rapid fire bolt operation sitting

regular shape. The black is, of course, the bullseye and counts 5. This wide space below the bullseye and narrow ring all counts 4; this other narrow ring is 3, and outside of that 2. Due to the irregularity of shape and complicated dimensions, and the fact that no sight changes are made between shots, there is no necessity to learn the dimensions of this target. After you have fired a string and it has been spotted, there is plenty of time to refer to your score book or a coach for the sight changes required to place your shot group in the bullseye the next time you fire at this same range. You will note from the location of the bullseye and 4 space that it is better to keep your shots low on this target for a low shot, if on the target, is a 4; whereas, if you are high, only a few will take you into the 3 or 2 ring.

58. Instructors carry out balance of lesson with their classes.

(1) Explain: The rapid-fire position is the same as the slow-fire position at this range. Previous to firing a string at 200 yards rapid fire, take the sitting position, make yourself comfortable and note position where the buttocks rest; stand up, without changing position of the feet, and hold the position until the targets appear. When the targets appear, drop into the marked spot, obtaining a predetermined comfortable position without waste of time or necessity of moving around to become comfortably located. In dropping into this position the right hand may or may not be used to break the fall. In this position, the butt of the rifle is kept at the shoulder and both elbows remain against the knees while operating the bolt; the eyes watching the target. Illustration 36.

(2) Demonstrates the assuming of the position and the firing of a string.

(3) Give four strings, simulating 200 yards rapid fire.

COMMON ERRORS

(*a*) Not picking position previous to actual firing.
(*b*) Slow in dropping into position.
(*c*) Not keeping eye on target while operating the bolt.
(*d*) Dropping rifle from shoulder to load from magazine.
(*e*) Jerking trigger due to rapidity of fire.
(*f*) The errors of the slow fire sitting position.

(4) Quizzes class:

Pertinent questions used in Sitting Position, lesson 8.

Questions used in 500 Rapid, lesson 10.

Q. Why should a trial sitting position be taken before the regular string is fired?

A. To save time dropping into a comfortable position after the targets appear.

Q. Is there any difference between the rapid fire and the slow fire sitting positions?

A. None.

Q. Should the piece be taken from the shoulder in operating the bolt?

A. No.

Q. Should either arm be removed from the leg in operating the bolt?

A. No.

Q. Can the trigger be pulled in this type of fire?

A. No.

Q. What factor has the most bearing on making a good score?

A. Correct trigger squeeze.

QUESTIONS ON TARGET DIMENSIONS

Q. Why is a knowledge of target dimensions necessary?

A. So that the firer may know how far his shots are from the center of the bullseye and thus make correct sight changes for the next shot or string.

Q. What is the diameter of the bullseye, A target?

A. Ten inches.

Q. Approximately, what is the width of the 4 ring?

A. Nine inches.

Q. The 3 ring?

A. Nine inches.

Q. What is the A target size?

A. Four by six feet.

Q. What is the bullseye diameter of the B target?

A. Twenty inches.

Q. What is the width of the 4 ring?

A. Nine inches.

Q. Of the 3 ring?

A. Nine inches.

Q. What is the size of the B target?

A. Six by six feet.

Q. What is an easy way of remembering the A and B scoring dimensions?

A. A working rule. Legal age is 21. The bullseye on the B target is 1 inch less than this figure, or 20 inches. Bullseye of A target and rings of both targets are, roughly, 11 inches smaller than the B bullseye, or 9 inches.

Q. Why are no dimensions given for the rapid-fire target?

A. Because the dimensions are complicated and hard to remember; also changes are not made for each shot; once firing is begun it is completed without a sight change.

Q. Then how are proper elevations determined?

A. After a string is fired and the group spotted, there is plenty of time to figure out the necessary changes by reference to score book or consulting the coach.

(4) Questions by class.

(5) Give three strings simulating rapid fire, 500 yards and 300 yards.

LESSON 13

O'CLOCK OF HITS—EQUIPMENT USED

B TARGETS AND SPOTTERS

LECTURE BY SENIOR INSTRUCTOR

59. To allow a shooter and his coach to speak of hits being made on the target and understand each other, a simple scheme for designating hits is used, called the clock system. A knowledge of this system is needed in order that hits may be designated without waste of words, and the exact location of these hits be easily comprehended. This is also true regarding the direction from which the wind blows. The wind deflects the bullet from its course. This deflection is overcome by use of the wind gage. The same strength wind blowing from a different angle has a different effect upon the bullet, and demands a

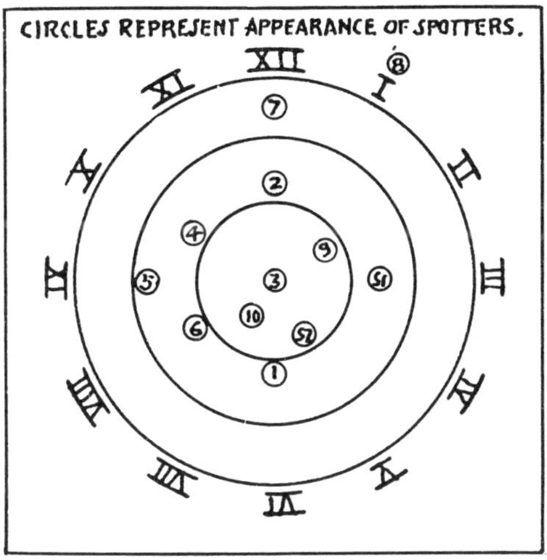

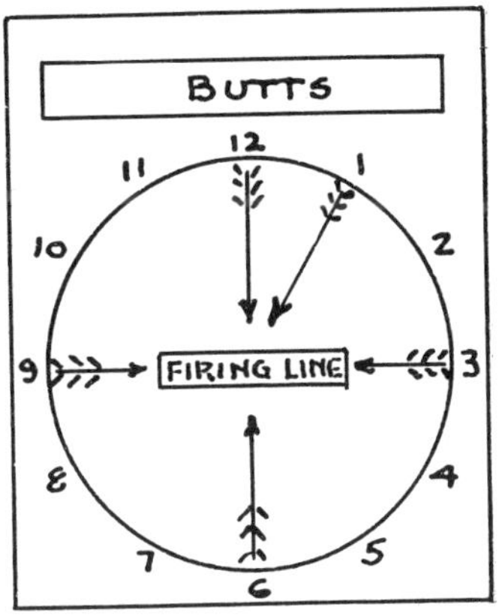

corresponding difference in the setting of the wind gage. To refer to winds from different angles necessitates some system of designation, and a knowledge of this system by the shooter. When you face the target, imagine a clock of which you are the center and the target 12 o'clock. Your right arm, if raised sideways, points to 3 o'clock; your left arm, raised, to 9 o'clock. A wind blowing from the target to you is called a 12 o'clock wind; if directly from your right, it is a 3 o'clock wind; if directly from your rear it is a 6 o'clock wind. Directions in between would be designated by other appropriate hours. Now imagine this target as a clock face, with the bullseye as the center; the top 12 o'clock; the right side 3 o'clock; the bottom 6 o'clock, and the left side 9 o'clock. With this knowledge it is easy to designate where a shot struck. If it struck here in the 4 ring, directly to the right, it would be a 3 o'clock 4; if straight up in the 3 ring, a 12 o'clock 3; if it struck the bullseye at the bottom, a 6 o'clock 5. If in the 4 ring, nearer 3 o'clock than 12 o'clock, a 2 o'clock 4, etc.

60. Instructors carry out balance of lesson with their classes.

(1) Holds school on o'clock of wind; giving example. Point out directions and give sufficient schooling to assure its being understood by the class.

(2) Using target, place spotter in various positions and require location and value to be given. Examples: Q. What is this? A. 10 o'clock 3. Q. This? A. 5 o'clock 4. Q. This? A. 11 o'clock 2. Q. This? A. 4 o'clock 3. Numerous and varied questions should be asked; the instructor assuring himself that the class thoroughly understands the subject.

(3) Simulate 3 strings, rapid fire 200 yards sitting, and 3 strings, rapid fire 300 yards prone. Check for all errors and correct where necessary.

(4) Questions by class.

LESSON 14

WINDAGE RULE; QUARTER POINT RULE; SQUARE RULE—EQUIPMENT USED

B TARGETS AND SPOTTERS

LECTURE BY SENIOR INSTRUCTOR

61. This lesson deals with sightsetting, calculation of the amount of windage to be used, and of the changes to be made on the sight to land your shots in the bullseye (demonstrate). Each line on this wind gage is one point of wind. Moving it to the right throws the bullet to the right; movement to the left, throws the bullet to the left. These lines are elevations, and the number below each line indicates the elevation reading for that line. Raising the elevation, raises the bullet strike; lowering the elevation drops the bullet strike. A wind from the right will blow the bullet to the left; the distance depending upon the force and direction of the wind. Before firing the first shot at any range, the wind gage must be set to offset the effect of the wind. There is a simple rule, called the **windage rule,** for determining windage. Multiply the force of the wind by the hundred unit in the range and divide the result by 10; this will give the windage required in quarter points. Example: 500 yard range, 10-mile wind blowing; 10 by 5 equals 5 quarter points, or 1¼ points. If the wind was blowing from the right, it would be 1¼ right windage. If the wind direction was from 2 or 4 o'clock, or 8 or 10 o'clock, a little less windage would be required—about a strong point. If the wind should be from 1 or 5, 7 or 11, about one-half of the computed windage would be required. The windages smaller than computed are required because the wind has most effect when blowing at right angles to the line of fire. As it changes around the effect becomes less; until it blows directly up or down the range, when it has no effect upon the bullet, i.e., no windage is required.

NOTE.—This rule is for 1906, 150 grain ammunition. If firing M–1, 172 grain ammunition, make your rule: Range times wind force divided by 13, which will equal the number of quarter points of windage required. The wind has only about three-quarters of the effect on the M–1, 172 grain ammunition that it has on the 1906 ammunition.

Frequently your shots will strike high or low and it is then necessary for you to know how to calculate the amount of change required in your elevation to place the next shot in the bullseye. There is a simple rule, called the **square rule**, for doing this. Squaring the unit representing hundredths in the range yardage will give the number of inches that a 100-yard change in your sight will produce on the target. Example: Firing at 600 yards, raise the elevation 100 yards and the bullet strike will be raised 6 by 6, or 36 inches on the target. Or, in other words, to move the bullet strike 36 inches on the target requires a 100-yard change in sight setting. To apply this rule, suppose you are firing at 500 yards and the bullet strikes 20 inches above the center of the bullseye; you know you must lower the sight, but how much? According to the rule, a 100-yard change at this range will move the bullet 25 inches; therefore, the amount to change the sight is twenty twenty-fifths or four-fifths of 100 yards or 80 yards. Besides having shots high and low, they will strike to the left and right, and you will wish to know how much change on the wind gage reading is necessary to place the next shot in the bullseye. For this, we have the **quarter point rule.** At any range, a change of one-quarter point on the wind gage will move the bullet strike on the target the same number of inches as there are hundreds in the range at which you are firing. Example: Firing at 500 yards, a quarter point change on the wind gage will move the bullet strike 5 inches on the target. At 600 yards, 6 inches; at 1,000 yards, 10 inches. Without a working knowledge of these rules, a shooter is at sea; his corrections will be pure guesswork and many points will be thrown away in an endeavor to land in the bullseye by guessing at the corrections to be made on the sights.

62. Instructors carry out balance of lesson with their classes.

(1) Assuming different directions and forces of winds for various distances, give problems and require the students to work them out. Review the windage rule, if necessary. Some men may never seem to grasp this rule, particularly if mentally slow.

(2) Give three strings, rapid fire 200 yards.

(3) School on quarter point and square rules. Place spotters in B target at various points and require sights to be properly set for the next shot.

(4) Give three strings, rapid fire at 500 yards.

(5) Questions by class.

LESSON 15

KEEPING THE SCORE BOOK—EQUIPMENT USED

B AND D TARGETS, SPOTTERS AND MARINE CORPS SCORE BOOKS

LECTURE BY SENIOR INSTRUCTOR

63. Score books are for the purpose of recording your shooting data. By this is meant a complete history of each string you fire; the kind of ammunition used, the time of day, direction and strength of wind, whether sunshine or cloudy light, windage and elevation used, plotting the call of each shot and where it actually struck the target, the value of each shot, etc. There are blank spaces in the book for all necessary data and they should be carefully filled out. Keeping this data serves several valuable purposes and breeds the habit of taking pains and observing, which is invaluable to a shooter. A study of the data kept frequently shows the shooter errors which he has made and helps him to avoid them in the future. Different ammunition sometimes requires a different elevation at the same range. Some men, due to their own peculiarities of aiming, etc., require more, or less windage when shooting rapid fire than they use at slow fire at the same range with the same strength and direction of wind. Some men, due to peculiarities of eyesight may use a different elevation at the same range for morning and afternoon light, or shooting in sunshine and clouds. The score book, being a written record, covers all these things and leaves nothing to memory, which often proves false. Before shooting at any range, a study of records made before at the same range will help in avoiding errors and getting better scores.

64. Instructors carry out balance of lesson with their classes.

(1) Explains a specimen score sheet as follows: The rifle number should be entered in order to be sure that you have not picked up the wrong rifle; the place and event are not necessary, except as a matter of interest in the future; the date should be entered as it may be necessary to refer to what you did on a certain day. The light, wind, mirage, direction of the sun, help to determine the zero of the rifle. The sight settings, hour, and weather conditions are of help in setting the sight correctly when again firing at that range. The most important phase of the score book is to plot the call, that is where the line of sight was when the rifle discharged. Plot this, and when the shot is signaled plot it. Compare the two, and if consistently apart you will know that your call is incorrect due to flinching; or, if you are not flinching, it will show errors in the rifle, ammunition or sight setting. The plotting of shots shows shot groups and what changes are needed. The score sheet should show a complete record of all that transpires while you are firing. The zero of your rifle, as determined from firing, must be shown. Don't trust

your memory. The zero is the amount of error on the sight for the range you are firing. For instance, if shooting at 500 yards and there is no wind blowing, yet your sight reading is 500 and one-half left windage, with your shots going into the center of the bullseye, then you have a one-half left wind zero and this must be considered when setting the sight for actual wind blowing.

(2) Using a B target and spotter, range 600 yards, assume hits for a string of slow fire at this range, announce data for each shot, require changes of windage and elevation to be worked out and each man fill in a score sheet. Check entries made. With D target, assume 10 hits rapid fire and require score sheet to be filled out.

(3) Ten minutes, simulate fire, 200 yards standing. Three strings, 200 rapid. Three strings, 300 rapid.

(4) Quizzes class:

Q. When should entries be made in score book?

A. Immediately as things occur for which entry should be made.

Q. If it is seen that a group of shots, as plotted in the score book, are above the center of the bullseye, what should be done?

A. The next time at the range, use less elevation. Use a change that would bring the center of the group to the center of the bullseye.

Q. What is the zero of a rifle?

A. That reading of the windage that has to be used when there is no wind blowing. If the wind gage in this case reads one-half left, that is called the zero and must be considered when setting the sights to allow for wind.

Q. Why should the kind of ammunition used be entered in the score book?

A. Different kinds of ammunition may call for different elevations at the same range. FA 1916 may call for 50 yards less elevation at 600 yards than US 1918, and it is important to know this.

Q. Why should the hour be shown?

A. To some men a morning and afternoon elevation at the same range may be different and it is well to know such differences for future firing.

Q. Why should the zero be known?

A. So that it will not be forgotten, and will be used in applying actual wind corrections.

Q. Why should wind and o'clock of wind be shown?

A. To help in figuring out windage the next time the same range is fired.

Q. What is mirage?

A. Apparent movement of air sometimes seen in looking at the target, when the sun shines or in some lights.

Q. Why an entry regarding mirage?

A. When mirage is heavy, it affects elevation for some men.

Q. Why an entry for regarding light?

A. To some men changes in light call for a change in elevation.

Q. Why an entry regarding the "call of the shot"?

A. To check the accuracy of the call against the actual position of the shot. If the call and the actual position of the shot are consistently apart, it denotes incorrect call due to flinching; or if assured the call is correct, errors of the rifle, ammunition or sightsetting.

Q. In brief, what entries should be made in the score book?

A. A statement of the conditions surrounding the firing of each shot which could effect the location of the shot.

Q. When should the "call" be entered?

A. The minute the shot is fired.

Q. When is the shot location and value entered?

A. The minute it is signalled from the pits.

Q. What are the figures on the left and bottom of some score sheets?

A. Elevation and wind gage corrections.

(5) Questions by class.

LESSON 16

ESTIMATING WIND; MARKING SCORING—EQUIPMENT USED

B AND D TARGETS, SPOTTERS, SCORE BOARD, MARKING DISKS (IF AVAILABLE)

LECTURE BY SENIOR INSTRUCTOR

65. Correct estimation of the wind is necessary in order to have the first shot of a string good, and to make correct windage changes during the firing of the balance of the string, if the wind shifts directions or varies in force. The ability to correctly estimate the wind comes from practice; there are a few guides. One is mirage, usually seen through a telescope or field glasses. When these wavy, watery heat waves or lines are seen to move slowly across the target, it indicates a slow wind—about 4 to 8 miles per hour. If rising straight up, no wind. If they appear to race and are close to the ground, it indicates a wind of over 10 miles per hour. Only practice will teach the force of the wind. Only mirage seen near or in the direction of the targets should be used in the estimation of wind. Wind may also be judged by the bend of trees, bushes, grass, etc. Moving smoke, flags blowing out, etc., can also be used. The nearer the target these indications, the more valuable they are, for it is wind near the target which has the most effect on the bullet's flight. Sand, paper, or grass thrown into the air may be used, and are good in determination of wind directions, but it must be remembered that there may be a difference in wind force and direction between the firing point and the targets.

Next, all men, while on the range, are called upon to perform two duties having no bearing on their own shooting. These are marking targets and keeping score for other shooters. When you are shooting some man will be marking your target and another man will be keeping your official score. You will expect both of these men to be right on the job. The scorer must be watchful and correctly record the value of the shot so that you can get full credit for what you make. The marker must give quick, accurate service on the target. He must see the shot holes immediately as the bullet strikes the target, use the correct marking disk in signalling the value of the shot, and place the center of the marking disk over the shot hole so that you can see exactly where your bullet struck. When marking or scoring for another man, give him the same service that you expect to get. If shots are marked with the wrong disk, or incorrectly recorded by the scorer, the shooter is cheated. If shots are carelessly disked, the shooter may be misled as to the exact location of his shots. If the marking and scoring is not efficiently carried out, it not only causes mistakes, but is irritating to the shooter and slows up the firing.

66. Instructors carry out balance of lesson with their classes.

(1) With a B target, D target, spotters, and marking disk explain marking targets (using B target). When a shot strikes in the bullseye, or in the 4 ring but nicks the edge of the bullseye, it is a 5. You place the peg of this white spotter in the hole in order that the firer may know just where his shot struck; run up the target and place the center of this white disk directly over the spotter to indicate the value of the shot to the firing line. Expose the disk for a second or two. If the shot struck in any other section of the target, a spotter is used and the appropriate marking disk for the correct value of the shot. This red disk indicates a 4; this black and white disk a 3; this black disk a 2. If the target is pulled and careful examination fails to show a hit, this red flag is slowly waved twice across the target face to indicate a miss. When the target is pulled down, the shot spotter is removed from the old hole, placed in the new (if appropriate for the last shot), and the old bullet hole pasted up. In rapid fire, spotters should be placed in all holes, in order that the firer may know where his hits are and determine the center of the group. It is good practice to jot down on paper the hits made, say: 4–5's; 3–4's; 2–3's; 1 miss, thus no mistakes will be made in disking the shots. After running the target up, indicate with the correct disk 4–5's; 3–4's; 2–3's; and 1 miss. In marking targets, the question often arises, Is it a 5 or a 4? A 4 or a 3? This, due to the bullet striking on the dividing line. Remember, if the bullet actually cuts the edge of the dividing line, you give the greater value. Always watch the target for hits and pull it immediately after it is struck. There are other orders applying to marking and conduct in the pits which will be read to you when you get actual pit duty.

(2) Give three strings, rapid fire 500 yards.

(3) School on scoring. (Use score board.) Explain: The firers' surnames, followed by their rank, are entered on the board in the order in which they are to shoot. Watch your man; when he fires, if the target is not promptly withdrawn, call to the telephone man, "Mark target ----------." When the target is disked, record the shot value in the proper square, and call, loud enough for the firer to hear plainly, "Corporal Jones, first-second shot a 5–4", or whatever the value may be. In rapid fire, the procedure is, "Corporal Jones, 1–5; 2–5's; 3–5's; 1–4; 2–4's; and so on." Usually in rapid fire scoring, the odd targets are scored first so that the adjacent even number scorers doing the calling make it possible for the odd number scorers to place the record on the board. When the even numbered targets are scored, the scorers change around—odd numbers calling; even numbers recording. If you do not clearly see a disk, or a manifest error is made on disking, in either rapid fire or slow fire, call to the telephone operator, "Redisk target ----------." Then the target will be redisked. A hit on the wrong target is scored a miss. Never erase the scores until directed by the range officer to do so. These are the principal points to be observed in scoring, although most ranges have special orders covering this duty which are published before men are actually assigned to duty as scorers.

(4) Give one 10-minute period, simulate 200 yards slow fire. Give one 10-minute period, simulate 600 yards sandbag rest.

(5) Quizzes class:

Q. In disking a shot, how should the disk be used?

A. Center of disk placed in front of shot hole.

Q. In scoring, why is the shooter's name and value of the shot called out?

A. So that the firer can detect mistakes and call attention to them in time to have them remedied.

Q. What disk is used to indicate a 5, 4, 3, 2, miss?

A. White, red, white and black, black, red flag.

Q. What is the result of using the wrong disk?

A. The firer is cheated, or receives more than he is entitled to.

Q. What is the result of not drawing the target down promptly when it is fired on?

A. The firer may think he has missed it and worry; delays due to this are irritating, and anything that worries or irritates a man who is firing is apt to hurt his score.

Q. How are shots marked in rapid fire?

A. Fives, then fours, then threes, etc.

Q. Why is a spotter used?

A. To indicate the exact position of the hit.

Q. Why is this necessary?

A. So that correct sight changes for the next shot may be calculated.

Q. Why disk the shot when spotters are used?

A. Because the scorer and firer could not tell fours from threes, threes from twos.

Q. If a shot strikes in the 3 ring, but cuts the black line marking the 4 ring, what is the value disked?

A. A four.

Q. If a target is not withdrawn promptly after it is fired upon, what should the scorer do?

A. Call, "Mark target __________."

Q. If the scorer is in doubt about a shot value, what should he do?

A. Call, "Redisk target __________."

Q. Why is fast work in the pits a good thing?

A. It gets the work done quicker and eliminates the possibility of the firer becoming irritated at unnecessary delays.

Q. Can you give other effects that fast pit work has on the firer?

A. Yes. Fast target operation reacts upon the firer and speeds him up in his work.

Q. What can happen if a scorer is unattentive?

A. Miss a disked shot and thereby mix up the recorded score.

Q. What else?

A. Cause confusion in that the firer will claim to have fired 10 shots, whereas the scorer will have recorded only 9.

(6) Questions by class.

LESSON 17

CARE AND CLEANING OF RIFLE WHILE ON RANGE—RANGE RULES—EQUIPMENT USED

CLEANING GEAR: CLEANING ROD, WIRE BRUSH, CLEANING PATCHES, CLEANING FLUID

LECTURE BY SENIOR INSTRUCTOR

67. Accurate shooting requires a clean, polished rifle bore and nothing will so injure its accuracy as improper or lack of care in cleaning. During range practice and for 7 to 10 days thereafter, utmost care should be given the bore. The bore should be carefully cleaned at the conclusion of each day's shooting, with the preparations provided, and daily for 7 to 10 days after range practice is completed. The object of cleaning is to remove the fouling deposited in the bore by firing. Of this fouling, there are three kinds: A deposit from the burnt primer, one from the burnt powder, and the metallic scrappings

1538—36——5

from the bullet as it passes through the barrel. A clean, polished bore does not rust as easily or as quickly as a dirty, fouled bore. Rust is the enemy of an accurate bore. If the bore is not kept clean, rust will form under the fouling and oil will have no effect upon it. A rifle with rust pits in it fouls much more easily than one without, besides being harder to keep clean. A rifle seldom actually wears out. Ninety percent of all rifles with inaccurate barrels become that way through improper cleaning. If a rifle is cleaned properly, it will stand thousands of rounds of firing without becoming unserviceable from firing.

Before you go on the range, it is appropriate that you be warned regarding Range Safety Regulations. Certain safety regulations are in force on all ranges; they must be observed. Their purpose is to prevent yourself or others from being injured in an accident, and to prevent your being subjected to certain disciplinary action as the cause of a serious accident. Bear in mind that it is **almost always a so-called "unloaded weapon"** which causes the accident.

Range safety precautions.—(1) No rifle shall be loaded except at the firing point, and then only upon the orders of the range officer in charge.

(2) When at the firing point with a loaded rifle the muzzle will be kept pointed in the direction of the target until the arm is either discharged or unloaded.

(3) When not at the firing point and in the act of firing, rifles shall be kept at all times with the bolts open.

(4) Never point a rifle at anyone you do not intend to shoot, nor in a direction where an accidental discharge may do harm. When on the range, do not snap for practice when in rear of the firing line.

68. Instructors carry out balance of lesson with their classes.

(1) Demonstrate cleaning a rifle.

Explain: Immediately after you finish firing for the day, go to the cleaning rack and clean your rifle in this fashion: Remove the bolt; wet the brass brush with powder solvent; run the brush through the barrel, from the bolt end out through the muzzle and then back. Do this three times if using a new brush and more if using an old brush. Then wet a clean patch in the solvent and run it through the barrel in a like manner, 10 times both ways. Repeat with a second patch. Then wipe out the barrel with a clean, dry patch. After this, inspect the barrel for cleanliness. If the bore is not bright or seems to be rough, repeat the above cleaning until it seems to be bright and smooth. After this, run a clean patch, soaked in solvent, through the bore one round trip. Remove solvent from muzzle and magazine. The rifle can now be put away for the night. Before firing in the morning, run a couple of clean, dry patches through the barrel to remove the solvent, and wipe off the bolt. This wire brush, which is

.32 caliber, is first used to break up and loosen any deposit which may be in the barrel. The following patches are to scrub out the fouling. The so-called "powder solvent" is used to loosen up and dissolve the fouling. It is left in the barrel overnight to further loosen and dissolve any fouling which may be present after cleaning; the oil which is in it will also prevent the formation of rust overnight. This solvent, however, cannot be used to replace oil as a preservative, as it will evaporate to a great extent in about 24 hours. So, if you are going to miss shooting for, say, 2 days, you should, after cleaning your rifle, run oil rather than solvent through the barrel.

NOTE.—If solvent is not available, clean as given in Basic Field Manual, chapter 1, Rifle Marksmanship.

(2) Hold school on the following range rules:

When on the rifle range always keep the bolt open when not on the firing line. Be careful not to load the rifle until on the firing line and about to fire. Before leaving the firing line, unload chamber and magazine and open bolt. For simulated fire (snapping in) take a position either on the firing line or its extension, and abreast of the firers. Be careful not to point or aim a rifle except at a target, and then only when on the firing line. Carry the rifle with muzzle elevated. These are all rules intended to prevent accidents. Keep rifles in the rack provided, when not in use. This rule is intended to preclude the careless laying of rifles on the ground, or setting them against post, benches, etc., where they may be walked upon or knocked down, with the likelihood of their being broken or having the blackening brushed off the sights. Keep the cartridges clean. Sand or grit on the cartridges will act like emery on the bore of your rifle. Remember, do not dig holes in the firing line; it has to be used by other shooters. Pay attention to the scorer when he calls the value of your shots, and check your score on the score board before leaving the line. The intention of these rules is to permit the immediate correction of errors; later correction will be difficult. Observe these range rules; it may save you much future grief.

(3) Quizzes class:

Q. What is the object in cleaning the rifle?

A. To remove fouling deposited in the bore.

Q. When should it be cleaned?

A. After the completion of shooting each day, and for 3 days after range practice is finished.

Q. Why should the rifle be cleaned each day for 3 days after range practice is finished?

A. To assure that all fouling is removed.

Q. What would happen if it was not?

A. Rust would form under the fouling.

Q. What object is there in having a clean, polished bore?

A. Makes it harder for rust to form.

Q. What is the disadvantage of a rusty bore?

A. Ruins accuracy, fouls easier, and is harder to clean.

Q. Why should the bolt of the rifle be kept open at all times on the range, unless shooting?

A. To avoid possibility of accident.

Q. Why should the ammunition be kept clean?

A. If dirty, the cartridge may stick in the chamber. Also, if sand is carried through the barrel by the bullet, it will wear the bore out as though sandpapered.

Q. What do you think the effect of rust upon metal is?

A. It wears the metal away.

Q. Can rust form overnight?

A. Yes.

Q. Why is the bristle brush used in cleaning?

A. To break the fouling, giving the solvent a better chance to do its work.

Q. What effect has the solvent on fouling?

A. It dissolves and washes it away.

Q. Which end is the rifle cleaned from?

A. The breech or chamber.

Q. Why is solvent left in the barrel overnight?

A. To work on any small particles of fouling which may be left in the barrel after cleaning, and also, to prevent rust from forming during the night.

Q. How does the solvent prevent rust?

A. It has a small amount of oil in it.

Q. Is it good for 48 hours, as a rust preventative?

A. No.

Q. If the rifle is put away for 48 hours or longer, what should be done?

A. A coat of oil should be placed in the barrel.

Q. Before firing in the morning, what should be done with the rifle?

A. Clean out the solvent and wipe the working surface of the bolt with an oily rag.

Q. If after cleaning the rifle it is still fouled, what should be done?

A. If the bore is not bright or seems to be rough, repeat the cleaning process until it becomes bright and smooth.

Q. What ruins barrel accuracy?

A. A rusty bore.

Q. Under normal conditions, is there any excuse for a rifle becoming rusty?

A. No.

Q. Give one safety regulation.

A, Never load a rifle until on the firing line, ready to fire.

Q. Give another one.

A. Never point or aim a rifle except on the firing line; and then only at a target.

Q. Another.

A. Always keep the bolt open when not on the firing line.

Q. Another.

A. When through firing, unload and leave bolt open.

Q. Should a rifle be laid on the ground or against a bench?

A. No.

Q. Why?

A. It may be stepped on or knocked over and broken.

Q. When should a score be checked?

A. Before leaving the firing line.

Q. Why not dig holes in the firing line?

A. Different men may not desire, or be able to use the same holes. The practice would then lead to a mass of holes on the firing line, with no one able to get into a comfortable position.

(4) Questions by class.

FINAL EXAMINATION

69. Before proceeding to .22-caliber practice, or instruction practice if no .22-caliber practice is held, all men should be given a final examination on the preliminary lessons. Those found deficient in this examination should be given such additional preliminary instruction as may be necessary to insure their qualification. The examination can be based upon the questions in the preceding lessons, and particular attention should be paid to those points in which errors are not readily discernible when viewing the men in simulated firing. In other words, men who now take the correct positions at simulated fire, can now be assumed to know the correct positions. Base the quiz mainly upon the intangibles; aiming, trigger squeeze, breathing while aiming, quarter point and square rules, etc.

Note.—If no .22-caliber rifles are available, and preliminary training is to be directly followed by .30-caliber practice, a great deal of simulated fire should be interspersed with the .30-caliber firing. This is to take the place of the preliminary training which is ordinarily continued in the .22-caliber firing.

.22-CALIBER SECTION—SMALL BORE

70. The preliminary training, when completed, should be followed by .22-caliber training. This small-bore training is, in short, an employment of the principles set forth in the preceding lessons. This application of these principles in actual firing maintains interest and affords the instructor an opportunity for the detection of errors which are apparent only in real firing. With the greatly reduced .30-caliber ammunition allowance, it permits of practice in those forms of fire which require considerable instruction for the attainment of proficiency, and which could not, otherwise, be had.

71. Efforts should be made to organize teams and promote small-bore matches. This stimulates and maintains interst in marksmanship, at the same time providing healthy amusement and diversion for the men.

LESSON 18

SMALL-BORE SHOOTING—EQUIPMENT USED

.22-caliber rifles, 50 feet, .22-caliber range, .22-caliber long rifle cartridges.

NOTE.—See Schedules part III.

LECTURE BY SENIOR INSTRUCTOR

72. We are now about to begin small bore, or .22-caliber practice. This involves the same principles that you have previously been taught. It is more interesting than simulating fire and gives a concrete example of your proficiency. It furnishes a check on your progress and gives your instructors (now coaches) the opportunity to detect the errors that you may be committing and to help you in overcoming them. It is your first opportunity to actually apply what you have learned, and the results will show how well or how poorly you have done so. This is known as the **U. S. Rifle, caliber .22, M1922 M1 (or M1922 M2)**. It is approximately the same weight as the .30-caliber rifle and, as closely as possible, was made to be a duplicate of that rifle in order that the two would be interchangeable for training purposes. It fires the .22 long rifle cartridge. This rear sight is known as the Lyman sight. Its graduations are entirely different from those on your .30-caliber rifle; the elevation being in minutes. Due to the fact that many men must use the same rifle, necessitating sight changes for each man, all firing will be done for group. By this I mean you will fire a 10-shot string, always aiming at the same spot, but making no sight changes to place your shots in the bullseye. The pattern that these 10 shots make is called the shot group. If it is small, the shooting is good. If large or scattered, you have committed errors. With a .22-caliber rifle, this method is quite as good as shooting for actual score, and saves having to learn to use a different sight and the application of the changes to keep each shot in the bullseye. It is not your score that counts, it is the size of the group, which is an indication of your proficiency or the commission of errors. This rifle is cleaned in the same manner as the .30-

caliber rifle. Each coach cleans the rifle used by this group, so you have no worries on that point. You can obtain the utmost good from this small-bore firing if you apply yourself to it, for it will fix in your minds the correct positions, coordinate and make a habit of those steps essential to the finished rifleman; namely, aim, take up slack, hold breath, squeeze the trigger, and call the shot. In the same, manner, this practice carelessly undertaken will handicap your .30-caliber shooting. The lack of recoil or noise to the discharge of this rifle will tend to make you careless in regard to your position and sling adjustment. This must be guarded against. The bolt operation is easier due to the difference in the cartridge and can lead to improper position and faulty bolt operation habits for the .30-caliber rifle. To offset this, all small bore rapid fire will be followed by simulating fire with the service rifle.

NOTE.—In shooting for group, approximate scores may be assigned and a record kept by using a transparent marking target to determine scores.

73. Instructors carry out balance of lesson with their classes.

(1) Sights in rifle to place group near the center of target. (B–1 target, prone position.) Explain to class: this rifle is sighted in for me. Due to variations in eyesight, different men, even though aiming correctly, will not have a group well-centered in the bullseye. However, the excellence of your shooting will not be judged by score, but by the size of the group made by your 10 shots. If your group can be covered by a dime at this range, your shooting is good. The smaller the group, the better your shooting. Always aim at 6 o'clock.

(2) Require all men to fire 10 shots, prone position. Take plenty of time. Correct visible errors in position, sling adjustment, etc. Watch for signs of flinching, or trigger jerking. Have men call each shot and note number of bad shots called in each string. **A too large group may mean** flinching, not squeezing trigger (not prevalent in small-bore shooting); not aiming correctly, variations in manner of aiming; not aiming at 6 o'clock each time; or breathing while firing.

(3) Go over each man's group with him. Mark name, date, and type of fire on the target for future comparison, or use scoring target, assign score, and keep a record of it.

(4) Continue balance of small-bore schedule for this day.

Cautions.—Watch for and correct all tendencies to carelessness. Each string of small-bore rapid fire should be followed by a similar, simulated string of .30 caliber rapid fire. This to counteract any bad habits which may be formed due to the entirely different bolt action on the .22-caliber rifle.

INDEX

Part II.—PISTOL MARKSMANSHIP

QUESTIONS AND ANSWERS

PART II.—PISTOL MARKSMANSHIP

1. When time permits, record practice with the rifle should be completed before commencing instruction and practice with the pistol. When time is limited and there is not sufficient time to conduct pistol practice after record practice with the rifle, pistol practice should be conducted concurrently with .22 caliber rifle practice. Pistol practice should not be conducted concurrently with .30 caliber rifle practice. Although certain features are similar for both rifle and pistol training there should be no divided interest when firing the rifle. Rifle and pistol training will be assimilated better if conducted separately.

2. In cases where men have not had preliminary training with the rifle it will be necessary to give preliminary training in sighting and aiming and trigger squeeze with the pistol. The training in trigger squeeze should be thorough.

LESSON 1

EQUIPMENT—PISTOLS AND TABLES

3. (**a**) **Lecture.**—The rifle is the basic weapon of the Marine Corps, yet duties are so varied that no man may know when his duties may require that he be armed with the pistol. When so armed he must know how to use the pistol effectively, and what is also important, he must know how to use it without danger to his own men, bystanders, and himself. Men regularly armed with the pistol are required to fire the pistol dismounted course each year. All others are required to fire the pistol dismounted course the first year of each enlistment and the pistol short course each remaining year of each enlistment. All men in the Marine Corps must be proficient in the use of the pistol, dismounting, assembling, care and cleaning, and must know the safety devices and tests of safety devices. During the actual firing of the pistol each man will be required to load the magazine, load the pistol, clear all jams, etc., himself, under the observation of the coach. He will be given no assistance from the coach in these steps, consequently he must be familiar with all details before he fires the first shot.

(**b**) **Lecture.**—The coach gives the following lecture and demonstrates each step: This is the automatic pistol (Colt), caliber .45, model 1911. Now, before proceeding further, I want to impress upon you one very important point. When you pick up a pistol for any purpose execute "unload." Take the position of raise pistol, press in the magazine catch, remove the magazine, pull back the slide and look in the chamber, release the slide, pull the trigger, and insert an empty magazine. Now we know the pistol is unloaded. When not actually in use all pistols on the range will have the chamber open with slide stop engaged.

FUNCTIONING OF THE PISTOL

Lecture.—The pistol is recoil operated. The force of the recoil drives the slide and barrel back together. The downward movement of the barrel then unlocks it from the slide, which continues the backward movement alone, opening the breech, cocking the hammer, and compressing the recoil spring. During the backward movement the empty shell is extracted by the extractor and ejected by the ejector. The slide having reached its rearmost position, the pressure of the recoil spring drives it forward again. During this return movement the slide carries a loaded cartridge from the magazine into the chamber. The hammer remains cocked and the pistol is ready to fire. The trigger pressure must be released after each shot in order that the trigger may re-engage. Due to recoil during firing this release of pressure on the trigger requires no effort on the part of the firer. When the magazine has been emptied the magazine follower pushes up the slide stop, thus locking the slide in the open position. This serves to remind the firer that the last shot has been fired.

DISMOUNTING AND ASSEMBLING—NOMENCLATURE

Lecture.—I will now dismount and assemble the pistol giving the nomenclature as I proceed. To dismount the pistol remove the magazine by pressing in the magazine catch; press the plug inward and turn the barrel bushing to the right until the plug is released from the barrel bushing, keep the thumb on the plug to prevent it flying out and striking you or being lost. Draw the slide rearward until the smaller recess in its lower left edge stands above the projection on the thumb piece of the slide stop; press against the end of the slide stop pin which protrudes from the right side of the receiver and remove the slide stop. Draw the slide forward from the receiver. The slide carries with it the barrel bushing, barrel, recoil spring guide, recoil spring, and plug. Lift out recoil spring guide; remove recoil spring and plug; turn barrel bushing to left and remove from slide; draw barrel forward from slide; push out link pin and remove link.

The pistol is now ready for cleaning, and this is as far as it needs to be dismounted for normal care and cleaning. For complete cleaning after completion of range practice, or for replacing some unserviceable part further dismounting is necessary.

To dismount the slide group press the rear end of the firing pin forward until clear of the firing pin stop; press firing pin stop down and out of slide; remove firing pin, firing pin spring, and extractor.

To dismount the receiver group, cock the hammer and remove the safety lock by pushing on pin from right and pulling outward on thumb piece when safety lock is midway between upper and lower positions. Lower the hammer and remove hammer pin from left of receiver; remove hammer. Push out housing pin from right and draw mainspring housing down and out of receiver. Remove grip safety and sear spring. Push out sear pin from right to left and remove sear disconnector. Compress mainspring and push out mainspring cap pin; remove mainspring cap; mainspring; and housing pin retainer.

There are two designs of magazine catch. To remove this type press checkered left end inward until right end projects so far from right side of receiver that it may be rotated one-half turn releasing magazine catch lock from seat in receiver; remove magazine catch and contents from receiver; remove trigger.

With this design (the improved magazine catch lock) press the left end of magazine catch inward and with short leaf of sear spring turn magazine catch lock one-quarter turn to left and remove magazine catch and contents. This design can be recognized by the magazine catch lock which has a slotted head.

(**c**) The instructor assembles pistol in reverse order, explaining each step and giving nomenclature.

The instructor requires each man to dismount and assemble the pistol, and to give nomenclature as he proceeds. The instructor prompts the pupil but does not assist him otherwise. The instructor sees that no part is forced in dismounting or assembling. Parts may be damaged by undue force in dismounting and assembling.

(**d**) The instructor gives the following instructions for cleaning the pistol:

CARE AND CLEANING

Lecture.—It is essential that all parts of the pistol be kept cleaned and oiled to prevent jams. To clean the pistol after firing, remove the slide from the receiver and the barrel from the slide. Clean the bore of the pistol the same as you clean the bore of the rifle. Clean all other parts with oil and leave a thin film of oil on them. After completion of range practice dismount the pistol completely and clean all parts.

OPERATION

(**e**) The instructor gives the following lecture on method of operation, demonstrating each step:

Lecture.—The magazine may be loaded with any number of cartridges from 1 to 7. Five are used in all range practice. The magazine is placed in the handle where it is engaged and retained by the magazine catch. Never strike the magazine a sharp blow with the hand—this will spread the lips of the magazine and cause jams. The slide is drawn fully back and released, thus bringing the first cartridge into the chamber. If the slide is open, push down the slide stop. The hammer is then cocked and the pistol is ready for firing (demonstrate unload). If desired to load with 8 cartridges for field use, draw back the slide, engage slide stop, load cartridge in chamber by hand, release slide stop, turn safety lock to safe and insert loaded magazine. The pistol may be carried safely at full cock with the safety lock in the locked position. Also, it may be carried safely with hammer down on a loaded chamber. It is impossible to fire the pistol except by a full blow from the hammer. It is impossible for the firing pin to even touch the primer except from a blow from the hammer as it falls, consequently, the pistol is safe with the hammer down (demonstrate unload).

The instructor requires each pupil to load and unload the pistol, magazine loaded with five cartridges. He demonstrates lowering hammer on loaded chamber and requires each pupil to execute this step.

SAFETY DEVICES

(f) **Lecture.**—The pistol is provided with four safety devices; the safety lock (manually operated), the half-cock notch, the grip safety, and the disconnector. The safety lock locks the slide and locks the hammer in the cocked position. The disconnector (automatic) positively prevents the release of the hammer unless the slide and barrel are in the forward position safely interlocked; this device also controls the firing and prevents more than one shot following each pull of the trigger. The grip safety (automatic) at all times locks the trigger unless the handle is firmly grasped and the grip safety pressed in. The half-cock notch prevents the hammer's falling and striking the firing pin except when the hammer is fully cocked and the trigger pulled. Thus it can be seen that accidents with the pistol are due only to extreme carelessness; the pistol can be fired only when fully cocked, safety off, pistol grasped so that grip safety is pressed in, and the trigger pulled.

TEST OF SAFETY DEVICES

(g) **Lecture.**—A defective safety device is a dangerous device. To insure that all safety devices are functioning correctly they are tested as follows:

Test of safety lock: Cock the hammer and press the safety lock upward into the safe position. Grasp the stock so that the grip safety is pressed in and squeeze the trigger 3 or 4 times. If the hammer falls the safety lock is defective and must be replaced.

Test of grip safety.—Cock the hammer and, being careful not to depress the grip safety, squeeze the trigger 3 or 4 times. If the hammer falls the grip safety is defective and must be replaced.

Test of disconnector.—Shove the slide one-quarter inch to the rear; hold slide in that position and squeeze the trigger. Let the slide go forward, maintaining the pressure on the trigger. If the hammer falls the disconnector is worn on top and must be replaced. Pull the slide all the way to the rear and engage the slide stop. Squeeze the trigger and at the same time release the slide. The hammer should not fall. Release the pressure on the trigger and then squeeze it. The hammer should then fall.

Test of half-cock notch.—Draw back the hammer until the sear engages the half-cock notch and squeeze the trigger. If the hammer falls the hammer or sear must be replaced. Draw the hammer back nearly to full cock and then let it slip. It should fall only to half-cock.

Defective parts causing malfunctions of safety devices should not be repaired—they should be replaced.

Failure of grip safety to function correctly may be due to weak leaf on sear spring. In such case replace sear spring.

The instructor then requires each pupil to test all safety devices.

SAFETY PRECAUTIONS

(**h**) **Lecture.**—The instructor states safety precautions as follows:

(1) Execute unload every time the pistol is picked up for any purpose. Never trust your memory. Consider every pistol as loaded until you have proved it otherwise.

(2) Always unload the pistol if it is to be left where someone else may handle it.

(3) Always point the pistol up when snapping it after examination. Keep the hammer fully down when the pistol is not loaded.

(4) Never place the finger within the trigger guard until you intend to fire or snap for practice.

(5) Never point the pistol at anyone you do not intend to shoot, nor in a direction where an accidental discharge may do harm. On the range, do not snap for practice while back of the firing line. When on the range and not at the firing point and in the act of firing, pistols will have their magazines withdrawn and slides pulled back.

(6) Before loading the pistol, draw back the slide and look through the bore to see that it is free from obstruction.

(7) On the range do not insert a loaded magazine until the time for firing.

(8) Never turn around at the firing point while you hold a loaded pistol in your hand, because by so doing you may point it at the man alongside of you.

(9) On the range, do not load the pistol with a cartridge in the chamber until immediate use is anticipated. If there is any delay,

lock the pistol and only unlock it while extending the arm to fire. Do not lower the hammer on a loaded cartridge, lock the pistol. Lowering the hammer on a loaded cartridge is for emergency use only when 8 shots may be desired.

(10) In reducing a jam first remove the magazine.

(11) To remove a cartridge not fired first remove the magazine and then extract the cartridge not fired from the chamber by drawing back the slide.

(12) In campaign, when early use of the pistol is not foreseen, it should be carried with a fully loaded magazine in the socket, chamber empty, hammer down. When early use of the pistol is probable, it should be carried loaded and locked in the holster or hand. Extra magazines should, in campaign, be carried fully loaded.

(13) When the pistol is carried in the holster loaded, cocked and locked, the butt should be rotated away from the body when drawing the pistol in order to avoid displacing the safety lock.

(14) Safety devices should be tested frequently. A safety device is a danger device if it does not work when expected.

The instructor then questions each pupil on safety precautions to insure that he knows them.

LESSON 2

PRELIMINARY TRAINING—EQUIPMENT—PISTOLS AND L TARGETS

4. (**a**) **Lecture.**—The requirements for good pistol shooting are very simple. All you need to know and apply is the correct position, method of holding the pistol, alinement of sights, 6 o'clock hold (slightly different for some pistols), and trigger squeeze.

(**b**) Position. (See illustrations nos. 37 and 38.)

Lecture.—The body is faced left or nearly so from the line of fire, feet 12 to 18 inches apart, head and body erect, body perfectly balanced, elbow and wrist straight, pistol grasped firmly but not tensely, barrel of pistol as nearly in prolongation of wrist and forearm as possible. When the sights are alined the hold is directed at 6 o'clock on the bullseye by action of the shoulder muscles only (instructor demonstrates position as he describes it). Never fire with bent elbow and wrist, and never attempt to direct the pistol on the target by movements of the elbow and wrist; such methods result in very unsteady holding, quick and erratic movements of the pistol, trembling, and jerking the trigger.

(**c**) Holding the pistol. (See illustration no. 38.)

Lecture.—In holding the pistol the hand grasps the pistol well up on the stock, the grasp on the stock being firm but not tense, firm enough to prevent the pistol shifting in the hand from recoil. The

Illustration No. 37

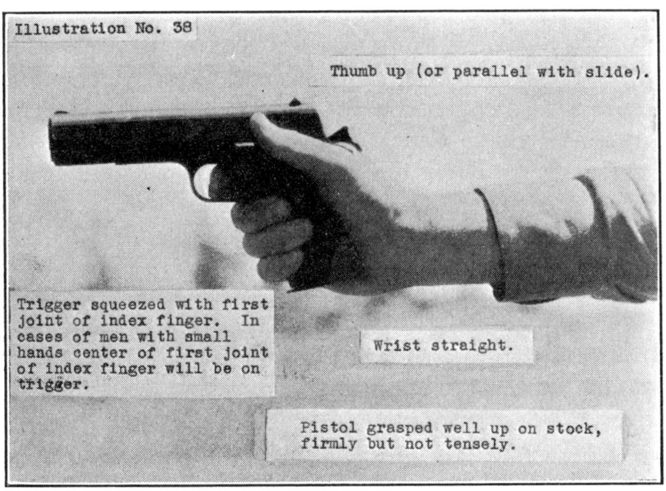

Illustration No. 38

thumb may be up, as indicated, or parallel with the slide, never down. The position with the thumb up is preferred by most good pistol shots.

(**d**) Trigger squeeze. (See illustration no. 38.)

Lecture.—The trigger is squeezed with the first joint of the index finger, the pressure being applied steadily the same as for the rifle. The second joint of the index finger can not be used to squeeze the trigger of the pistol. The pressure on the trigger is applied straight to the rear.

(**e**) Alinement of sights.

Lecture.—The sights are alined so that the top of the front sight is level with and in the center of the notch of the rear sight. The correct hold is at 6 o'clock at 15 yards and slightly lower at 25 yards. All pistols do not shoot exactly the same and it may be necessary to hold slightly right or left, above or below, the stated point of aim. Learn where the pistol shoots and correct your point of aim until the group is centered on the bullseye; do not attempt to correct by changing alinement of front and rear sights.

The above lectures (*a*) to (*e*) are given consecutively, the instructor demonstrating each step. The coach then requires all men to "snap in", simulating three scores slow fire. He corrects all errors in position, holding the pistol, and trigger squeeze.

(**f**) Rapid fire. Equipment: Pistols, L targets, piece of strong cord 4½ feet long for each man.

The cord is tied to thumb piece of hammer, the other end of cord being held in left hand of firer for cocking the pistol. The instructor demonstrates use of cord for cocking pistol when simulating rapid fire. The time will be 11 seconds for all simulated rapid fire. Practice for simulated timed fire is not given.

Lecture.—In rapid fire the trigger squeeze is the same as for slow fire except that the trigger squeeze is much faster. The time limit for timed fire is 15 seconds, for rapid fire 11 seconds. You have not more than 2 seconds to aline the sights, pick up the bullseye, and squeeze the trigger. Start the trigger squeeze as soon as you pick up the sights and get on the target; coordinate your trigger squeeze and holding so that the shot is fired as soon as your hold is correct, or approximately so, at 6 o'clock; keep your eye on the bullseye; keep your arm and wrist straight and avoid flourishes and unnecessary movements between shots. Above all, squeeze the trigger. Keep your arm and wrist straight and the pistol will naturally return to approximately correct alinement of sights and position after each shot.

The instructor then requires all men to "snap in" three scores, simulating rapid fire. He corrects all errors.

5. A great amount of "snapping in", both slow fire and rapid fire, under close supervision, is required to produce good results in pistol marksmanship training. The amount of this practice will depend upon the time available for pistol marksmanship training. A minimum of 6 hours is required. During this training, note especially the trigger squeeze, as this is the all-important point in pistol shooting. Also look for good "follow through" after trigger squeeze; that is, the pistol should not move or jump as the hammer falls. If the pistol does not move it indicates the pupil is relaxing when he believes the pistol should fire, his method of grasping the pistol is incorrect, or he is jerking the trigger. In incorrect method of grasping the pistol or attempting to get the index finger far enough through the trigger guard to squeeze the trigger with the second joint will cause the pistol to jump as the hammer falls.

PISTOL RANGE PRACTICE

6. (**a**) Conduct "snapping in", both slow fire and rapid fire, concurrently with range practice firing.

(**b**) Detail one coach per target for instruction at firing point during range practice firing. Correct all errors.

(**c**) Enforce all safety precautions.

(**d**) Require the firer to load magazine, load pistol, clear all jams and malfunctions himself without any physical assistance from the coach. Require each man to dismount and clean the pistol he uses.

(**e**) Do not permit men to "aim in" before targets appear in timed and rapid fire, or before the silhouette target is turned for each shot in quick fire. This defeats the purpose of training in rapid and quick fire. Require them to remain at "raise pistol" until the targets appear.

(**f**) The ammunition allowance for the dismounted course is 250 rounds per man, and should be used for range practice and record as follows:

	Rounds
First day:	
Demonstration by coach	20
Practice	45
Second day: Practice	45
Third day: Practice	45
Fourth day: Practice	50
Fifth day: Record	45
Total	250

Where time is not available for preliminary pistol training prior to the week during which pistol range practice is conducted all pistol training and firing will be conducted during the week of .22 caliber rifle practice as stated in appended schedules. Where time is available for pistol preliminary training and range firing after completion of .30 caliber range firing the preliminary training will be given as directed by the range officer, but the practice firing will be as stated in the schedules.

PRELIMINARY INSTRUCTION

1. QUESTIONS AND ANSWERS

Q. Should pistol practice be conducted at same time with .30 caliber rifle firing?

A. No. It should follow rifle training. If there is not sufficient time after rifle training, pistol training should be held the week before .30-caliber firing. There should be no divided interest when firing the rifle.

Q. Why is pistol training necessary?

A. Any man may be armed with the pistol at some time. He must know how to use it effectively and safely.

Q. Does the coach assist the pupil in loading and unloading the pistol during firing?

A. No. The pupil must learn all steps necessary in handling and firing the pistol before he fires the first shot.

Q. What is the most important safety precaution in the use of a pistol?

A. Always take the position of "raise pistol" and execute "unload" when you pick up a pistol for any purpose.

Q. How is functioning of the pistol taught?

A. By lecture and demonstration by the instructor.

Q. State, in your own words, the functioning of the pistol.

Q. How is dismounting, assembling, and nomenclature taught?

A. By lecture and demonstration by the coach or instructor. The pupil learns the steps rapidly when dismounting, assembling, and giving nomenclature concurrently. It is also taught by requiring the pupil to dismount and clean the pistol he uses.

Q. How is the pistol cleaned?

A. Use the same methods used in cleaning the rifle.

Q. What dismounting is required for cleaning the pistol during range practice?

A. Remove slide and dismount slide group.

Q. How many cartridges are loaded in the magazine for range practice?

A. Five.

Q. How is a loaded pistol carried?

A. If immediate use is not anticipated it is carried with 7 cartridges in the magazine, chamber empty and hammer down. When immediate use is anticipated it is carried loaded with 7 cartridges, 1 in the chamber, hammer cocked and locked. Eight cartridges may be used but 7 is the prescribed number.

Q. How is the unloaded pistol carried?

A. With the hammer down.

Q. For safety, how are pistols carried on the range?

A. Magazine out, slide back, slide stop engaged. Load the pistol only when on the line and ready to fire.

Q. Is a loaded pistol safe with the hammer down?

A. Yes. With the hammer resting on the firing pin, the firing pin does not touch the primer. The firing pin can touch the primer only by a blow from the hammer as it falls, and the cartridge can be fired only by a full blow from the hammer.

Q. Name the four safety devices.

A. Safety lock, grip safety, disconnector, and half-cock notch.

Q. Which safety devices are automatic?

A. Grip safety, disconnector, and half-cock notch.

Q. What safety device is manually operated?

A. The safety lock.

Q. What is the function of each of the four safety devices?

A. The safety lock locks the hammer in the cocked position and at the same time locks the slide. The grip safety locks the trigger at all times until pressed in when the handle is grasped firmly. The disconnector prevents the release of the hammer unless the slide and barrel are all the way forward and safely interlocked, it also prevents more than one shot following each pull of the trigger. The half-cock notch prevents the hammer falling and striking the firing pin except when the hammer is fully cocked and the trigger pulled.

Q. What can you say about a defective safety device?

A. A defective safety device is a dangerous device. Safety devices must be checked frequently.

Q. State how you test each safety device.

Q. State the 14 safety precautions for use of the pistol.

Q. What publication is always available to all men for reference in use of the pistol?

A. The Marine Corps Score Book.

Q. What use should an instructor always make of the Marine Corps Score Book?

A. Assign certain pages for study prior to the day on which instruction will be given on this subject matter.

PRELIMINARY TRAINING

1. QUESTIONS AND ANSWERS

Q. Is pistol shooting difficult to learn?

A. No. It is very simple, all the firer needs to do is hold at 6 o'clock (slightly different with some pistols), and squeeze the trigger.

Q. What is necessary in training to be able to hold steadily.

A. First, the correct position; second, the correct methods of grasping the pistol; third, the correct alinement of sights.

Q. What is the correct position?

A. Faced left, or nearly so; feet 12 to 18 inches apart; head and body erect; body well balanced; elbow and wrist straight; pistol grasped firmly but not tensely; barrel as nearly in prolongation of the wrist and forearm as possible.

Q. How is the pistol directed at 6 o'clock on the bullseye?

A. By action of the shoulder muscles only, you point the pistol and arm as though it were one unit.

Q. What is the result of trying to shoot with bent elbow and wrist?

A. It should never be done. This causes trembling and unsteady holding, quick and eratic movements of the pistol and tends to produce flinching and jerking the trigger. It also slows down rapid fire by difficulty in picking up alinement of sights and getting back on target after each shot in rapid fire.

Q. How do you grasp the pistol?

A. Grasp the pistol well up on the stock, firmly but not tensely; the thumb may be up or parallel with the slide—never down.

Q. How do you aline the sights?

A. The top of the front sight is level with the top of and in the center of the notch in the rear sight.

Q. How do you squeeze the trigger?

A. With the first joint of the index finger. The increase of pressure should be steady, the same as for the rifle.

Q. What is the all-important point in pistol shooting?

A. Trigger squeeze.

Q. Can the second joint be used?

A. No. The finger is not long enough. You cannot get the second joint on the trigger except by holding the pistol between thumb and fore finger; and it is impossible to shoot the pistol accurately that way.

Q. What two things cause all failures to qualify as expert with the pistol?

A. Failure to squeeze the trigger and being slow and hesitant in rapid fire.

Q. How is position, correct holding of the pistol, alinement of sights, and trigger squeeze taught?

A. By slow fire practice at snapping in, carefully supervised by the coach.

Q. How do you detect incorrect trigger squeeze?

A. Watch the trigger finger and watch the pistol. If the trigger is squeezed properly the pistol will not move as the hammer falls. An incorrect method of grasping the pistol may make correct trigger squeeze impossible. First make sure the pistol is grasped correctly. To correct poor trigger squeeze the coach occasionally squeezes the trigger for the pupil, his fore finger over the pupils finger which is on the trigger.

Q. How is proficiency in steady holding and trigger squeeze developed?

A. By a great amount of snapping in before beginning range practice.

Q. How is proficiency in rapid fire obtained?

A. By practice snapping in rapid fire, the pistol being cocked by a cord tied to the thumb piece of the hammer. The cord is held in the left hand by the firer.

Q. What time limit is used in all rapid fire snapping in?

A. Eleven seconds.

RANGE PRACTICE

1. QUESTIONS AND ANSWERS

Q. What faults does the coach look for in correcting poor rapid fire?

A. Unnecessary flourishes with the pistol, slowness in getting back on the target, and slowness in picking up alinement of sights, slow trigger squeeze, and flinching or jerking the trigger. Of these faults, flinching and jerking the trigger is the worst.

Q. When do you start the trigger squeeze in rapid fire?

A. As soon as you get the sights alined and get on the target. The trigger squeeze and the holding must be so coordinated that the shot is fired as soon as the hold is correct at 6 o'clock.

Q. What is the true test of a good pistol shot?

A. Good rapid fire and quick fire scores. The pistol is an emergency weapon and to use it effectively one must be able to fire quickly and accurately. A good slow fire shot who cannot shoot well in rapid and quick fire is a poor pistol shot.

Q. During pistol range firing is snapping in dispensed with?

A. No.

Q. How many coaches are used on the firing line?

A. One per target.

Q. Does the coach clear a jam for the pupil?

A. No. He requires the pupil to clear the jam.

Q. What is the cause of most jams with the pistol?

A. Damaged magazines, caused by striking the magazine a sharp blow with the hand when loading it into the pistol.

Q. Is the firer permitted to aim in before the target appears in timed fire, rapid fire, or quick fire?

A. No. Remain at raise pistol until target appears.

Q. What is the best method of using the 250 rounds of ammunition allowance for pistol dismounted course?

A. Best results will be obtained by firing once over the course each day, the extra ammunition being used the last day before record for practice in rapid fire.

Q. How are safety precautions developed as fixed habits?

A. By enforcing them during range practice.

Part III.—SCHEDULES

SCHEDULE OF RECRUIT TRAINING, RIFLE MARKSMANSHIP—COURSE OF 3 WEEKS

FIRST WEEK

SCHOOL RANGE EQUIPMENT

(*a*) Shaker aiming device, one or more, complete.

(*b*) A target, on 1-inch boards, with holes for spotters.

(*c*) B target, on 1-inch boards, with holes for spotters.

(*d*) D target, on 1-inch boards, with holes for spotters.

(*e*) Target carrier with A, B, and D targets (for rapid-fire exercises).

(*f*) Spotters (ten 3-inch spotters, ten 5-inch spotters).

(*g*) Score board.

(*h*) Oversize representation of rear sight with movable drift slide and wind gage, for sight-setting instruction (any desirable size from 1 to 2 feet in width, constructed of beaver board or 1-foot boards).

(*i*) Practice dummy ammunition (or range dummy ammunition when safety factors permit use).

(*j*) Text used in United States Marine Corps Score Book (for use of men under instruction).

MONDAY

7:30 TO 11:30 A. M.

Lectures.—Nomenclature, care, and cleaning of the rifle; sighting and aiming; sight setting; blackened sights.

1 TO 4:30 P. M.

Exercises.—Sighting and aiming; sight setting.

When the recruit platoon arrives on the school range, it is divided into groups of equal numbers, one group to each coach on duty on the school range. Shooting coats are padded, rifles inspected, and score books issued before any preliminary training begins. This completed, the coaches proceed with the training in the sequence stated in the schedule. Coaches are assigned duties for which they are best qualified.

The group division is for the purpose of preventing the training from becoming monotonous, and for rotation of groups from one coach to another. The instruction is alternately exercises and lectures, with 10 minutes' rest period to each hour. This method is followed through all school range instruction.

Tuesday

7:30 TO 11:30 A. M.

Lectures.—Position and use of the sling; holding the breath; trigger squeeze; bolt operation; and rapid fire.

Exercises.—One hour: Review sighting and aiming; 3 hours: Positions, which includes the following in order stated:

(*a*) Blackening the sights.

(*b*) Use of the sling.

(*c*) Sighting and aiming (during snapping in).

(*d*) Taking up the slack.

(*e*) Holding the breath.

(*f*) Trigger squeeze.

(*g*) Positions both slow and rapid fire in following order: Prone, prone with sandbag, sitting, kneeling, standing, standing to prone, standing to sitting.

1 TO 4:30 P. M.

Lectures.—Same as A. M.

Exercises.—One hour: Trigger squeeze; 2½ hours: Positions and snapping in slow fire; bolt operation and rapid fire (use target carrier and D target, specified time limits).

Note.—Do not use bolt operation exercises as specified in Basic Field Manual (with trigger tied back). Bolt operation exercises are in simulating rapid fire only.

Wednesday

7:30 TO 11:30 A. M.

Lectures.—Zero of the rifle; the square rule; the windage rule; the quarter point rule; o'clock of hits; force and direction of winds; estimating wind. (Give exercises in above subjects during lecture periods.)

Exercises.—One hour: Review positions and trigger squeeze; 3 hours: Snapping in slow fire; bolt operation and rapid fire.

1 TO 4:30 P. M.

Lectures.—Same as A. M.

Exercises.—One hour: Trigger squeeze; 2½ hours: Snapping in slow fire; bolt operation and rapid fire.

Thursday

7:30 TO 11:30 A. M.

Lectures.—Same as Wednesday; and, keeping the score book; range duties and range rules. (Give exercises in lecture subject matter during lecture periods.)

Exercises.—Same as Wednesday.

1 TO 4:30 P. M.

Lectures.—Same as A. M.

Exercises.—Same as Wednesday.

Friday

7:30 TO 11:30 A. M.

Thorough review of week's work.

1 TO 4:30 P. M.

Note.—Progress of all men must be noted throughout week's training and backward men given special attention daily.

Saturday

7:30 TO 11:30 A. M.

Left vacant to make up for lost time due to weather or other causes.

SECOND WEEK

Divide platoon into 2 groups, 1 group on .22-caliber range, other on pistol range. Group assignments to .22-caliber range and pistol range are for one-half day periods (A. M. and P. M.), each group completing training listed each day.

Monday

7:30 A. M. TO 4:30 P. M.

Coaches sight in rifles to center shot groups on bullseye—no sight changes by firer permitted.

Firing conducted with object of obtaining small-shot groups consistently in same place. To be satisfactory, shot groups should be about the size of a dime for all positions, slow and rapid fire, except for standing position (representing 200 yards slow fire) and sitting and kneeling positions (representing 300 yards slow fire).

Object of .22-caliber firing is to develop correct sighting and aiming; correct positions and steady holding; taking up the slack, holding the breath, and trigger squeeze; ability to fire 10 shots in rapid fire well within the time limits and secure 10 good hits; calling the shot. The above essentials to good shooting should be fixed habits by completion of .22-caliber practice.

Rapid fire is conducted in 10 shot scores, specified time limits. Coach removes empty magazine, firer loads second magazine.

All tendency to snap shoot, jerk the trigger, or flinch should be eliminated during .22-caliber practice. Prevent these bad habits from being formed. It is more difficult to correct them than prevent them from being formed.

.22-caliber firing; once over course, plus extra score standing and extra score sitting and kneeling, 90 rounds.

Bolt operation and rapid-fire practice with dummy ammunition precedes and follows all .22-caliber firing. Snapping in slow fire (all positions) is given in conjunction with rapid-fire practice.

Note.—This training is necessary during .22-caliber firing to counteract easy bolt operation, short bolt throw, and better trigger pull on .22-caliber rifles. Detail one coach for this instruction alone.

Lectures and exercises.—Sight changes (elevation and windage) to center groups in slow and rapid fire (use A, B, and D targets with spotters). Require sight setting in conjunction with these exercises.

.22-caliber firing.—Once over course plus extra score standing and extra score sitting and kneeling, 90 rounds.

Pistol training.

Lectures and preliminary training.

Demonstration by coach, 20 rounds.

Pistol practice, 45 rounds.

TUESDAY

7:30 A. M. TO 4:30 P. M.

.22-caliber training.—Bolt operation and rapid fire; snapping in slow fire; lectures and exercises same as Monday.

Pistol training.

Lectures and preliminary training.

Pistol practice, 45 rounds.

WEDNESDAY

7:30 A. M. TO 4:30 P. M.

Same as Tuesday.

THURSDAY

7:30 A. M. TO 4:30 P. M.

Same as Tuesday.

FRIDAY

7:30 A. M. TO 4:30 P. M.

Pistol record fired Friday.

SATURDAY

7:30 A. M. TO 11:30 A. M.

Left vacant.

THIRD WEEK

1. Divide the platoon into 8 relays, 4 relays firing A. M., 4 relays firing P. M. Relays alternate for A. M. and P. M. firing. Detail one coach per target used.

2. Rapid-fire practice with .30-caliber ammunition at each rapid-fire range should be given with mixed ammunition; that is, dummy cartridges interspersed with live cartridges. It should be done by firing the first strings with 2 live cartridges and 8 dummies. The number of live cartridges may be increased each string until the firer is sufficiently trained to fire full strings with live cartridges. This permits many strings of rapid fire with a minimum of expenditure of ammunition. Backward firers may be continued on the first stage until marked improvement is shown. In all cases it is conducted exactly the same as though it were record firing. In using 2 live

cartridges, 1 live cartridge should be placed at random in each clip. This should be done by the coach in order that the firer may not know which are live cartridges. In the strings with 5 and 5 of each type interspersed, 2 dummies should be used in one clip and 3 in the other. The number of mixed strings that can be fired will depend upon the ammunition available. This type of practice is invaluable for rapid-fire training. Rapid fire tends for some reason to incline men to flinch. The use of dummies interspersed with live cartridges enables the coach to detect this and also to demonstrate to the firer that he is flinching. If the cartridges are mixed so that he will not know which is which, it trains him to squeeze his shots in rapid fire. This type of practice must be conducted with care, otherwise it develops a tendency on the part of the firer to watch each cartridge as it is loaded into the magazine endeavoring to spot the live ammunition and then give extra care to the firing of the live cartridge and giving the dummies careless mechanical attention. Before firing the first mixed string, instructors should give a short talk to the effect that—

the first rapid fire strings which you will fire will be strings in which live cartridges will be mixed with dummy cartridges. These clips will be loaded by the coach in order that you will not know which are dummies and which are live cartridges. The object of this is to train you to be good rapid-fire shots. For some reason rapid fire has a tendency to develop flinching in the firer. If this habit is formed it will ruin your firing. The firing of these mixed strings develops the mechanics of rapid fire and counteracts the tendency to flinch. The firing of these mixed strings is for your benefit; do not endeavor to ascertain which cartridges are dummies and which are live ones. Devote the same attention to the firing of each cartridge in the string.

Monday

A. M. AND P. M.

Slow fire, sighting in.
200 yards—5 shots prone, target A.
300 yards—5 shots prone, target A.
500 yards—5 shots prone, target B.
600 yards—5 shots prone w/sandbag, target B.
Total, 20 rounds.

Tuesday

A. M. AND P. M.

200 yards—5 shots, slow fire, standing, target A.
200 yards—10 shots, rapid fire, target D.
300 yards—3 shots sitting, 2 shots kneeling, slow fire, target A.
300 yards—10 shots, rapid fire, target D.
500 yards—5 shots, slow fire, prone, target B.
500 yards—10 shots, rapid fire, target D.
600 yards—5 shots, slow fire, prone w/sandbag, target B.
Total, 50 rounds.

Note.—Rapid fire conducted as stated in paragraph 2.

Wednesday

A. M. AND P. M.

200 yards—5 shots, standing, target A.
200 yards—10 shots, rapid fire, target D.
300 yards—5 shots, slow fire, 3 sitting, 2 kneeling, target A.
300 yards—15 shots, rapid fire, target D.
500 yards—5 shots, slow fire, target B, prone.
500 yards—15 shots, rapid fire, target D.
600 yards—5 shots, slow fire, prone w/sandbag, target B.
Total, 60 rounds.

Note.—Rapid fire conducted as stated in paragraph 2.

Thursday

A. M. AND P. M.

Course A, less sighting shots at 600 yards.
Total, 70 rounds.

Friday

A. M. AND P. M.

Course A, record.
Total, 72 rounds.
Total ammunition used, 272 rounds.

SCHEDULE OF TRAINING, RIFLE MARKSMANSHIP—COURSE OF 3 WEEKS (FOR MEN FIRING FOR REQUALIFICATION)

FIRST WEEK

SCHOOL RANGE EQUIPMENT

(*a*) Shaker aiming device, one or more, complete.
(*b*) A target, on 1-inch boards, with holes for spotters.
(*c*) B target, on 1-inch boards, with holes for spotters.
(*d*) D target, on 1-inch boards, with holes for spotters.
(*e*) Target carrier with A, B, and D targets (for rapid-fire exercises).
(*f*) Spotters (ten 3-inch spotters, ten 5-inch spotters).
(*g*) Score board.
(*h*) Oversize representation of rear sight with movable drift slide and wind gage, for sight-setting instruction (any desirable size from 1 to 2 feet in width, constructed of beaver board or 1-foot boards).
(*i*) Practice dummy ammunition (or range dummy ammunition when safety factors permit use).

Monday

7:30 TO 11:30 A. M.—1 TO 4:30 P. M.

When firing detail arrives on the school range, it is divided into groups of equal numbers, one group to each instructor. Shooting coats are padded, rifles inspected, and score books issued before any preliminary training begins. This completed, the instructors proceed with the training in the sequence stated in the schedule. Instructors are assigned duties for which they are best qualified.

The group division is for the purpose of preventing the training becoming monotonous and for rotation of groups from one instructor to another. The instruction is alternately lectures and exercises, with 10-minute rest periods to each hour. This method is followed through all school range instruction.

Time: Rifle instruction, 6 hours; pistol instruction, 1½ hours.

Rifle

Lectures.—Nomenclature, care and cleaning of the rifle, sighting and aiming, sight setting, blackening sights.

Exercises.—Sighting and aiming, sight setting.

Pistol

Lectures.—Functioning of the pistol; dismounting and assembling; nomenclature; care and cleaning; operation; safety devices; test of safety devices; safety precautions.

Exercises.—Dismounting and assembling.

TUESDAY

7:30 TO 11:30 A. M.

Rifle

Lectures.—Positions and use of the sling; holding the breath; trigger squeeze; bolt operation and rapid fire.

Exercises.—One hour: Review sighting and aiming; 3 hours: Positions, which include the following in order stated:

(*a*) Blackening the sights.

(*b*) Use of the sling.

(*c*) Sighting and aiming (during snapping in).

(*d*) Taking up the slack.

(*e*) Holding the breath.

(*f*) Trigger squeeze.

(*g*) Positions, both slow and rapid fire in following order: Prone, prone with sandbag, sitting, kneeling, standing, standing to prone, standing to sitting.

1 P. M. TO 4:30 P. M.

Rifle

Lectures.—Same as A. M.

Exercises.—One-half hour: Trigger squeeze, 1½ hours: Position and snapping in, slow fire; bolt operation and rapid fire. (Use target carrier and D target, specified time limits.)

NOTE.—Do not use bolt operation exercise as specified in Basic Field Manual (with trigger tied back). Bolt operation exercises are in simulated rapid fire only.

Pistol

One and one-half hours.

Lectures.—Position; Holding the breath: trigger squeeze, alinement of sights; rapid fire.

Exercises.—Snapping in slow fire. Rapid fire exercises.

Wednesday

7:30 TO 11:30 A. M.

Rifle

Lectures.—Zero of the rifle; The Square Rule; The Windage Rule; The Quarterpoint Rule; O'clock of Hits; Force and Direction of Wind; Estimating Wind (give exercises in above subjects during lecture periods).

Exercises.—One hour: Review positions and trigger squeeze; 3 hours: Snapping in slow fire; bolt operation and rapid fire.

1 TO 4:30 P. M.

Rifle

Lectures.—Same as A. M.

Exercises.—One-half hour: Trigger squeeze; 1½ hours: Snapping in slow fire, bolt operation and rapid fire.

Pistol

One and one-half hours.

Same as Tuesday, special attention to correction of errors.

Thursday

7:30 TO 11:30 A. M.

Rifle

Lectures.—Same as Wednesday; and keeping the score book; range duties and range rules.

Exercises.—Same as Wednesday.

1 TO 4:30 P. M.

Same as Wednesday.

Friday

7:30 TO 11:30 A. M.

Thorough review of week's work.

1 TO 4:30 P. M.

NOTE.—Progress of all men must be noted throughout week's training and backward men given special attention daily.

Saturday

7:30 TO 11:30 A. M.

Left vacant to make up for lost time due to weather or other causes.

SECOND WEEK

Divide the firing detail into 2 groups, 1 on the pistol range and 1 on the .22-caliber range. When a group completes firing on the .22-caliber range, it is sent to the pistol range, the group on the pistol range goes to the .22-caliber range.

Automatic rifle practice, rifle grenade practice, and hand grenade practice is held from 1 p. m. to 4:30 p. m., grenade practice following automatic rifle practice.

.22-caliber firing is conducted with the object of obtaining small shot groups consistently in the same place for successive scores, development of correct sighting and aiming, positions, trigger squeeze, steady holding, and ability to fire all 10 shots in rapid fire well within the time limits. The coach removes the empty magazine, the firer loads the second magazine in rapid fire.

To counteract easy bolt action, short bolt throw, and better trigger pull on .22-caliber rifles, all .22-caliber firing will be preceded and followed with snapping in with the service rifle and rapid-fire exercises with dummy ammunition.

Monday

7:30 TO 11:30 A. M.

Rifle

Coaches sight in .22-caliber rifles to center shot groups on bullseye. No sight changes permitted by firer.

Lectures.—Sight changes in elevation and windage to center shot groups in slow and rapid fire. Use A, B, and D targets with spotters. Require sight setting in conjunction with these lectures.

Exercises.—Sight setting; positions and snapping in slow fire; rapid-fire exercises with dummy ammunition.

.22-caliber firing.—Once over .22-caliber course, 70 rounds.

Pistol

Exercises.—Snapping in, slow fire. Rapid-fire exercises.

Pistol-practice firing.—Once over course D, 45 rounds.

1 TO 4:30 P. M.

Browning automatic rifle, fire 1,000-inch course. Complete rifle grenade instruction and practice.

Tuesday

7:30 TO 11:30 A. M.

Rifle

Same as Monday.

Pistol

Same as Monday.

1 TO 4:30 P. M.

Browning automatic rifle, instruction practice, course A, table I. Hand grenade practice.

Wednesday

7:30 TO 11:30 A. M.

Rifle

Same as Monday.

Pistol

Same as Monday.

1 TO 4:30 P. M.

Browning automatic rifle, instruction practice, course A, table II. Hand grenade practice.

Thursday

7:30 TO 11:30 A. M.

Rifle

Same as Monday.

Pistol

Fire short course. Other pistol training same as Monday.

1 TO 4:30 P. M.

Browning automatic rifle, instruction practice, course A, table III. Hand grenade record practice.

Friday

7:30 TO 11:30 A. M.

Rifle

Same as Monday.

Pistol

Record firing, course D.

1 TO 4:30 P. M.

Browning automatic rifle, record practice, course A.

Saturday

7:30 TO 11:30 A. M.

Left vacant.

THIRD WEEK

1. Divide the detail into 8 relays, 4 relays firing A. M., 4 relays firing P. M. Relays alternate for A. M. and P. M. firing. Detail one coach per target used.

2. Rapid-fire practice with .30-caliber ammunition at each rapid-fire range should be given with mixed ammunition; that is, dummy cartridges interspersed with live cartridges. It should be done by firing the first strings with 2 live cartridges and 8 dummies. The number of live cartridges may be increased each string until the firer is sufficiently trained to fire full strings with live cartridges. This permits many strings of rapid fire with a minimum expenditure of ammunition. Backward firers may be continued on the first stage until marked improvement is shown. In all cases it is conducted exactly the same as though it were record firing. In using 2 live cartridges, 1 live cartridge should be placed at random in each clip. This should be done by the coach in order that the firer may not know which are live cartridges. In the strings with 5 and 5 of each type interspersed, 2 dummies should be used in one clip and 3 in the other. The number of mixed strings that can be fired will depend upon the ammunition available. This type of practice is invaluable for rapid-fire training. Rapid fire tends for some reason to incline men to flinch. The use of dummies interspersed with live cartridges enables the coach to detect this and also to demonstrate to the firer that he is flinching. If the cartridges are mixed so that he will not know which is which, it trains him to squeeze his shots in rapid fire. This type of practice must be conducted with care, otherwise it develops a tendency on the part of the firer to watch each cartridge as it is loaded into the magazine endeavoring to spot the live ammunition and then give extra care to the firing of the live cartridge and giving the dummies careless mechanical attention. Before firing the first mixed string, instructors should give a short talk to the effect that—

the first rapid fire strings which you will fire will be strings in which live cartridges will be mixed with dummy cartridges. These clips will be loaded by the coach in order that you will not know which are dummies and which are live cartridges. The object of this is to train you to be good rapid-fire shots. For some reason rapid fire has a tendency to develop flinching in the firer. If this habit is formed, it will ruin your firing. The firing of these mixed strings develops the mechanics of rapid fire and counteracts the tendency to flinch. The firing of these mixed strings is for your benefit; do not endeavor to ascertain which cartridges are dummies and which are live ones. Devote the same attention to the firing of each cartridge in the string.

1538—36——7

Monday

A. M. AND P. M.

Lecture.—Care and cleaning of rifle.
Slow fire, sighting in.
200 yards—5 shots prone, target A.
300 yards—5 shots prone, target A.
500 yards—5 shots prone, target B.
600 yards—5 shots prone w/sandbag, target B.
Total, 20 rounds.

Tuesday

A. M. AND P. M.

200 yards—5 shots, standing, target A.
200 yards—5 shots, rapid fire, target D.
300 yards—3 shots sitting, 2 kneeling, slow fire, target A.
300 yards—10 shots, rapid fire, target D.
500 yards—5 shots, slow fire, prone, target B.
500 yards—10 shots, rapid fire, target D.
600 yards—5 shots, slow fire, prone w/sandbag, target B.
Total, 45 rounds.

NOTE.—Rapid fire conducted as stated in paragraph 2.

Wednesday

A. M. AND P. M.

Course A, less sighting shots at 600 yards.
Total, 70 rounds.

Thursday

A. M. AND P. M.

Course A, record.
Total, 72 rounds.
Total ammunition used, 207 rounds.

Friday

Left vacant.

Saturday

Left vacant.

SCHEDULE OF TRAINING, RIFLE MARKSMANSHIP—COURSE OF 2 WEEKS (FOR MEN WITH TWO OR MORE QUALIFICATIONS AS EXPERT OR SHARPSHOOTER)

FIRST WEEK

When the number of men in the firing detail is sufficient to warrant division of the detail, divide the detail into 2 groups, 1 on the .22-caliber range and 1 on the pistol range. When a group completes firing on the .22-caliber range it is sent to the pistol range, the other group to the .22-caliber range. All .22-caliber and pistol firing is completed each day by 11:30 a. m.

Automatic-rifle practice, rifle-grenade practice and hand-grenade practice is held from 1 p. m. to 4:30 p. m., grenade practice following automatic-rifle practice.

Only men who consistently qualify as expert riflemen will be assigned as coaches, preferably noncommissioned officers or medal winners in Division and Marine Corps competitions. In shooting ability they should excel the expert riflemen whom they coach.

.22-caliber firing is conducted with the object of obtaining small shot groups consistently in same place, for successive scores, development of correct sighting and aiming, positions, trigger squeeze, and ability to fire all 10 shots in rapid fire well within the time limits. Coach removes empty magazine, firer loads second magazine in rapid fire.

To counteract easy bolt operation, short bolt throw, and better trigger pull on .22-caliber rifles, all .22-caliber firing will be preceded and followed with snapping in with the service rifle and rapid fire exercises with dummy ammunition.

MONDAY

7:30 TO 11:30 A. M.

Rifle

Inspect and repair rifles.

Pad shooting coats.

Issue score books.

Coaches sight in .22-caliber rifles to center shot groups on bullseye. No sight changes permitted by firer.

Lectures.—Brief lectures on nomenclature. Care and cleaning, blackening sights, sighting and aiming, sight setting, positions and use of sling, trigger squeeze, and rapid fire.

Exercises.—Sight setting, positions and snapping in slow fire, rapid-fire exercises with dummy ammunition.

.22-caliber firing.—Once over .22-caliber course, 70 rounds.

Pistol

Lectures.—Brief lectures on functioning, operation, safety devices, test of safety devices, safety precautions, position, holding the pistol, trigger squeeze.

Exercises.—Snapping in slow fire. Rapid-fire exercises.

Pistol-practice firing.—Once over course D, 45 rounds.

1 TO 4:30 P. M.

Browning automatic rifle, fire 1,000-inch course. Complete rifle grenade instruction and practice.

TUESDAY

7:30 TO 11:30 A. M.

Rifle

Lecture.—Zero of the rifle; the square rule; the windage rule; the quarterpoint rule; o'clock of hits.

Exercises.—Sight changes; changes in windage and elevation to center groups on bullseye in slow and rapid fire.

.22-caliber firing.—Once over .22-caliber course, 70 rounds.

Pistol

Exercises.—Snapping in, slow fire. Rapid fire exercises.

Pistol-practice firing.—Once over course D, 45 rounds.

1 TO 4:30 P. M.

Browning automatic rifle, instruction practice, course A, table I. Hand-grenade practice.

WEDNESDAY

7:30 TO 11:30 A. M.

Rifle

Lectures.—Force and direction of winds; estimating wind.

Exercises.—Changes in windage and elevation to center shot groups on bullseye in slow and rapid fire; estimating wind.

.22-caliber firing.—Once over .22-caliber course, 70 rounds.

Pistol

Same as Tuesday.

1 TO 4:30 P. M.

Browning automatic rifle, instruction practice, course A, table II. Hand grenade practice.

Thursday

7:30 TO 11:30 A. M.

Rifle

Same as Wednesday.

Pistol

Fire short course.

Exercises.—Snapping in, slow fire. Rapid-fire exercises.

Pistol-practice firing.—Once over course D, 45 rounds (plus extra ammunition for rapid fire).

1 TO 4:30 P. M.

Browning automatic rifle, instruction practice, course A, table III. Hand grenades, record practice.

Friday

7:30 TO 11:30 A. M.

Rifle

Same as Wednesday.

Pistol

Record firing, course D.

1 TO 4:30 P. M.

Browning automatic rifle, record practice, course A.

Saturday

7:30 TO 11:30 A. M.

Left vacant.

SECOND WEEK

1. Divide the detail into 8 relays, 4 relays firing A. M., 4 relays firing P. M. Relays alternate for A. M. and P. M. firing. Detail one coach per target used.

2. Rapid-fire practice with .30-caliber ammunition at each rapid-fire range should be given with mixed ammunition; that is, dummy cartridges interspersed with live cartridges. It should be done by firing the first strings with 2 live cartridges and 8 dummies. The number of live cartridges may be increased each string until the firer is sufficiently trained to fire full strings with live cartridges. This permits many strings of rapid fire with a minimum of expenditure of ammunition. Backward firers may be continued on the first stage until marked improvement is shown. In all cases it is conducted exactly the same as though it were record firing. In using 2 live cartridges, 1 live cartridge should be placed at random in each clip. This should be done by the coach in order that the firer may not

know which are live cartridges. In the strings with 5 and 5 of each type interspersed, 2 dummies should be used in 1 clip and 3 in the other. The number of mixed strings that can be fired will depend upon the ammunition available. This type of practice is invaluable for rapid-fire training. Raipd fire tends for some reason to incline men to flinch. The use of dummies interspersed with live cartridges enables the coach to detect this and also to demonstrate to the firer that he is flinching. If the cartridges are mixed so that he will not know which is which, it trains him to squeeze his shots in rapid fire. This type of practice must be conducted with care, otherwise it develops a tendency on the part of the firer to watch each cartridge as it is loaded into the magazine in an endeavor to spot the live ammunition and then give extra care to the firing of the live cartridge and giving the dummies careless mechanical attention. Before firing the first mixed strings, instructors should give a short talk to the effect that—

the first rapid-fire strings which you will fire will be strings in which live cartridges will be mixed with dummy cartridges. These clips will be loaded by the coach in order that you will not know which are dummies and which are live cartridges. The object of this is to train you to be good rapid-fire shots. For some reason rapid fire has a tendency to develop flinching in the firer. If this habit is formed, it will ruin your firing. The firing of these mixed strings develops the mechanics of rapid fire and counteracts the tendency to flinch. The firing of these mixed strings is for your benefit; do not endeavor to ascertain which cartridges are dummies and which are live ones. Devote the same attention to the firing of each cartridge in the string.

Monday

A. M. AND P. M.

Lecture.—Care and cleaning of rifle.
Slow fire, sighting in.
200 yards—5 shots prone, target A.
300 yards—5 shots prone, target A.
500 yards—5 shots prone, target B.
600 yards—5 shots prone w/sandbag, target B.
Total, 20 rounds.

Tuesday

A. M. AND P. M.

200 yards—5 shots, standing, target A.
200 yards—5 shots, rapid fire, target D.
300 yards—3 shots, sitting, 2 kneeling, slow fire, target A.
300 yards—10 shots, rapid fire, target D.
500 yards—5 shots, slow fire, prone, target B.
500 yards—10 shots, rapid fire, target D.
600 yards—5 shots, slow fire, prone, with sandbag, target B.
Total, 45 rounds.

Note.—Rapid fire conducted as stated in paragraph 2, page 4.

Wednesday

A. M. AND P. M.

Course A, less sighting shots at 600 yards.
Total, 70 rounds.

Thursday

A. M. AND P. M.

Course A, record.
Total, 72 rounds.
Total ammunition used, 207 rounds.

Friday

Left vacant.

Saturday

Left vacant.